COMBAT AMMUNITION

EVERYTHING YOU NEED TO KNOW

COMBAT AMMUNITION
EVERYTHING YOU NEED TO KNOW

DUNCAN LONG

Citadel Press
Secaucus, New Jersey

Published by Citadel Press
A division of Lyle Stuart, Inc.
120 Enterprise Ave., Secaucus, N.J. 07094
In Canada: Musson Book Company
A division of General Publishing Co. Limited
Don Mills, Ontario

Queries regarding rights and permissions should be addressed to: Lyle Stuart, 120 Enterprise Avenue, Secaucus, N.J. 07094

Manufactured in the United States of America

ISBN 0-8065-1043-9

Contents

Warning

Technical data presented here on handloading and on firearm use, adjustment, and alteration inevitably reflects the author's personal experience with equipment and components under specific circumstances which the reader cannot duplicate exactly. The information in this book should therefore be used for guidance only and approached with great caution. Neither the author nor the publisher assumes any responsibility for the use or misuse of information contained in this book.

Acknowledgments

I must extend my thanks to the many companies that so graciously loaned (or even gave) equipment and components to me for testing during the writing of this book. The improvements that have been made in ammunition design and quality control are truly amazing. American manufacturers should be very proud of the job they're doing.

A thank-you must go to Rose-Marie Strassberg and the other folks at Paladin Press for their help and encouragement in getting things together for this book.

Special thanks, too, to my dad, Paul F. Long, who helped at the last minute in printing photos for this book, and to Maggie and Kristen, who made it possible for me to have the extra time needed to finish this project.

1. What Do You Need in Combat?

You may someday face an enemy in mortal combat. Winning this battle will be all that counts. To win, you must have skill and a good firearm. Unfortunately, many fighters forget that the ammunition which feeds their weapon may be the deciding factor in whether they live or die.

When we look at ammunition, it is necessary first to know what the caliber of weapon you use can and cannot do. This is not as easy as one might hope. In a bit, we'll sort out the terminal ballistics for different rounds and how they go about doing their damage. You also have to be able to hit your target. While many shooters assume that misses are their own fault, in fact, many rounds have very limited accuracy. While long shots of up to 300 yards in combat are practical with rifles, other weapons are quite limited. Thirty yards is the maximum reliable range for a shotgun (except when slugs are used), and 100 yards for a pistol, with only a few yards being the normal combat range. Inherent inaccuracy, ballistic arcs, shot spread, air resistance to different bullet shapes, and even crosswinds all work against the shooter.

The bullet you use makes a world of difference as well. There are now a wide variety of bullet types available for most combat calibers. You must pick the one that suits your needs. For example, if you're shooting at an armed burglar in your city apartment, quite a different type of ammunition is needed from that used against an enemy soldier wearing body armor. During night fighting, you may need tracer ammunition and will want ammunition that puts out minimal muzzle flash. Body armor, weather conditions, and the combat environment can all make finding the right ammunition a very complex task.

You must also keep in mind what you want your ammunition to do. It's good to remember the differences between hunting and self-defense, especially in combat shooting where long ranges are involved. You don't aim to eat your enemy (I hope), so it doesn't make any difference if he can crawl away or even if he recovers weeks later. You only care that he is unable to continue fighting for a while, while with deer or other game, the goal is to drop an animal in its tracks. Another difference is that good hunters only take sure shots. In combat, on the other hand, enemies are often all but unseen in semiconcealed positions. Military studies suggest that at least with the rifle, a number of shots fired quickly may be more effective than one carefully aimed shot.

Another consideration is your opponent. A man's brain will greatly affect your bullet's effectiveness. Combatants with minor wounds may go into shock and die because they imagine the worst. Anger, fear, or the need for revenge can drive a wounded man to superhuman feats when, from a medical standpoint, he should be dead. Thus, hunting and combat resemble each other only in that firearms are involved in both.

There's a lot to consider. Let's look at some cold facts.

REAL-LIFE ENCOUNTERS

No two shootings in real-life encounters are alike. There are documented cases of burly guys falling like rocks when struck with .22 long rifle bullets and other times when criminals have taken off running despite chest wounds created by 12-gauge shotguns. Don't go by individual incidents or stories; they may be flukes rather than rules to live by when you're deciding which caliber would be ideal for combat situations.

STOPPING POWER

Bullets don't knock people over or down. If people were bowling pins, bullets would knock them over. But people are made of spongy, liquid-filled tissue, which absorbs a bullet's energy very effectively. Therefore, the term "knock-down power" is quite misleading; a better term is "stopping power." The exact mechanism of stopping power has long been misunderstood. Until recently, it was assumed that the more energy a bullet shed in a body, the greater was its stopping power; but this is only part of the story. The main consideration is the amount of damage the bullet does in shedding its energy.

Imagine two bullets of equal caliber and weight. One has a rounded point and the other has a hollow point that mushrooms. Both are propelled at the same speed and hit a living target at identical points. Both come to a stop in the bodies of the targets so that equal amounts of energy are lost. But the mushrooming nose of the hollow-point bullet creates a massive wound in a short distance while the rounded point of the other round tends to push flesh out of its way rather than severing or damaging it. It is easy to see that the mushrooming bullet will do a better job at stopping an assailant.

TESTING: FACTS AND FALLACIES

Actual shootings, however, are not done scientifically with identical bullets hitting all victims in the same place. Thus, testing is done on animals or ballistic gelatin. While much of the test data may be classified, it generally filters down by word of mouth or in actual documents.

A number of tests have been run during the last century to determine the best pistol caliber and ammunition/bullet types. One of the most influential of these was the Thompson-LaGarde series of 1904. Unfortunately, these tests were not done scientifically and the results were garbled in favor of the .45 round.

The exact procedures of the Thompson-LaGarde tests weren't commonly known until 1983. The evidence is startling. For the first phase of the test (which had the most weight with military planners several decades following the test), only thirteen cows of varying weight and age were used. Nine-millimeter and .38-caliber ammunition used in the test did not have bullets with expanding or flat points while some of the .45 calibers did. Weapons jammed, yet the tests continued to compare the recovery ability of cattle hit during rapid fire.

The second phase of the test involved testing ammunition on cadavers. It, too, was flawed, since a bullet's effectiveness was judged mainly by the movement its impact caused as noted by an observer, without measurements. While Xrays and dissection were also used, the main consideration was this subjective observed movement. Despite the lack of scientific procedure, these tests were accepted as gospel for almost eighty years.

Another old way of measuring bullet effectiveness is Julian S. Hatcher's formula, presented in his book *Pistols and Revolvers and Their Use* in 1927. The formula didn't agree with the Thomson-LaGarde tests, however, so in 1935, Hatcher revised it to support the erroneous test! A lot of people still recommend or choose pistol calibers for defense using this formula.

The first really scientific test of pistol bullets came in 1974 when medical examiner Vincent J. DiMaio published an article in the *FBI Law Enforcement Bulletin*. This article, entitled "A Comparison of Wounding Effects of Commercially Available Ammunition Suitable for Police Use," caused a lot of controversy, since it was based on tests done in a scientific manner using gelatin blocks as dense as human tissue, and it contradicted many of the assumptions about stopping power of the Hatcher formula.

This article was followed in 1975 by the Department of Justice's *An Evaluation of Police Handgun Ammunition*. This, too, used gelatin blocks for the test and proved that the .45 caliber was not the most effective round.

After finding that the Thompson-LaGarde tests were inaccurate and that the Hatcher formula was skewed in the wrong direction to support the conclusions of these tests, a new rating system was created by Carroll E. Peters for his book *Defensive Handgun Effectiveness*. In the 1980s, the Power

Index Rating system was developed by Edward A. Matunas. This system is also good at predicting the effectiveness of a round without extensive testing. We'll take a look at these formulas later and see how they can be used, or misused, in predicting the effectiveness of a pistol round in combat.

For rifle-caliber ammunition, probably the most influential work done after World War II was the U.S. Army's ALCLAD study. Though it set out to deal with the requirements for better body armor, ALCLAD also examined such elements as the range at which casualties occurred, what happened when the human body was struck by bullets or shell fragments, and the frequency and distribution of combat wounds. The study's conclusions, based on over three million casualty reports from the two world wars and the Korean War, flew in the face of much military doctrine. ALCLAD found that:

- nearly random shots produced greater casualties than aimed fire in combat;
- rifle fire was seldom used effectively at greater than 300-yard ranges;
- most rifle casualties were produced at ranges of 100 yards or less; and
- even expert marksmen could seldom hit targets beyond 300 yards because of terrain features or their own need to remain under cover.

COMBAT RANGES: HOW MUCH IS ENOUGH?

Long-range marksmanship is not as important in combat as many of us might think—with a combat rifle, the maximum need will probably be a 300-yard killing capability. (While many argue that you'll need more range in a desert or mountainous area, it should be noted that the Israelis switched from the .308 to the .223 round following their combat experience in the desert. Likewise, Afghan fighters tend to give up their long-range, bolt-action rifles for the new USSR .22-caliber assault rifles.)

A rifle chambered for .308 Winchester or the similar .30-06 will do the job, and then some. So will the .223, less recoil and weight of rifle and ammunition. Thus, from a combat-range standpoint, it makes sense to go with the .223, provided you can put up with some of its other limitations.

Ranges with shotguns are surprising in a different way: shotguns themselves dictate their range and that range is very short indeed, well under 300 yards. Though slugs will reach out to 100 yards, the shotgun is not rifled, nor do the slugs actually spin in flight, despite the rifled pattern on their sides. A good slug pattern at 100 yards is 10 inches.

With shot, the size and the bore's choke will determine the useful range. In general, the pattern of buckshot limits the range at which it can reliably inflict a disabling hit to 25 to 30 yards. The pattern can be made denser by using smaller shot, but smaller shot will lose more of its energy by the time it reaches an opponent, so again, the chances of a disabling wound are too small to guarantee your future. Though we'll look at some ways to overcome some of these problems in other chapters, it is good to bear in mind the shotgun's inherent limitations with commercial ammunition. (Heckler & Koch and Winchester are experimenting with ammunition for experimental shotguns, which is not usable in commercial weapons, which may increase the shotgun's range. Such improvements in ammo are still down the road, though.)

Pistols are generally used at very close ranges. While the pistol round is capable of accurate fire out to 100 yards and beyond in the hands of a skilled shooter, generally its actual maximum combat range is around 7 to 21 yards. With a pistol you need to be "fustest with the mostest." Accuracy, firepower, etc., all take a backseat to the round's ability to stop your opponent in his tracks before he can harm you. In many ways, the pistol has more need for stopping power than the rifle because of the extremely close ranges involved in most pistol shootings.

Pistol ammunition is also used in submachine guns. Generally, rounds fired from the submachine gun have better potential accuracy and greater muzzle velocity than the pistol. Rounds that are effective in pistols will be more effective in submachine guns, provided the weapon functions reliably with the round. Sixteen-inch-barreled carbines which fire pistol ammunition offer an even greater increase in velocity, so these, too, will do well with any ammunition that does well in pistols.

WOUNDING EFFECTS

The deciding factor in how effective a bullet will be in stopping an opponent is how much damage it will cause. There are a number of wounding forces, including punctures, hydrostatic shock, bullet fragmenting, multiple projectiles, vacuum/sonic shock, and projectile tumbling.

Puncture Wounds

Pistol bullets generally cause their wounding effects by puncturing. A number of methods have been developed to increase the damage done by a puncture wound. In order for a bullet to dump its energy and not exit through the shooting victim, it is necessary to increase the size of the puncture wound by increasing the bullet's diameter. (Such a bullet is less likely to overpenetrate walls or go through an assailant and wound a bystander.) If a person is limited by conventions of war or departmental rules to full-metal jacketed, or FMJ bullets (bullets which have a full metal shell over the tip), increasing the caliber of the round is the most practical way to increase its effectiveness. This is why the slow-moving .45 has been so much more effective than smaller-caliber rounds–it makes a bigger hole. That's its only plus, and that's all that is needed when a combatant is limited to FMJ bullets. (Since a rounded tip tends to push through tissue rather than damaging it, a bullet's wounding potential can also be increased by flattening its nose. This is not too practical with automatic loading pistols since it may cause functioning problems; it is a good way, though, to increase the effectiveness of revolver bullets.)

Another route with FMJ is to increase the round's velocity. This is a little iffy, however, in pistol ammunition. As we'll see when looking at rifle ammunition, pistol rounds would become very lethal if they moved at over 2,000 feet per second (fps), but that isn't a normal pistol velocity. Therefore, the fast-moving FMJ bullet gains lethality only when it chances to strike bone or body armor. This is why the .45 bullets generally move at rather slow speeds: faster speeds don't increase their wounding potential to any great extent and may actually decrease the safety of the bullet because of overpenetration.

Multiple Projectiles

Submachine guns and shotguns take advantage of multiple projectiles. It would seem that two bullets or pellets in an enemy would be twice as effective as one. In fact, they are (at least on paper) four times as deadly. Military studies show that the number of hits increases the likelihood of serious injury by the square of the number of projectiles. Thus, two hits are a whole lot better than one, and three–which is often cited as the ideal number with automatic weapons–do a theoretical damage nine times as great. Because of this, submachine guns using rounds like the .32 Auto, normally ineffective when single hits are produced, can be effective in the full-automatic mode. The shotgun is effective in close combat for much the same reason. Its projectiles spread very slightly within the 25-yard range, so that a well-placed hit will create a large number of puncture wounds that literally chew a hole in a shooting victim.

A new variation on the multiple-projectile round is the prefragmented Glaser Safety Slug, a hollow bullet filled with lead shot. The bullet penetrates the skin and then breaks into a number of projectiles which quickly spread to create multiple internal wounds. Actual shootings have proven that this round has the compounded lethality of other multiple projectiles; it is probably the most effective commercial bullet.

High-velocity rounds can create multiple projectiles from parts of the body as well as the bullet itself. For example, a rifle bullet striking bone will create a very massive wound compared to its effect when it passes through a boneless area. While a pistol bullet or shotgun pellet may not cause serious injury when hitting bone, an identical wound with a rifle bullet, however, will often be fatal. This is especially true with glancing head wounds or wounds to the rib cage. Rifle hits to arm or leg bones may be totally disabling for the same reason, since the fragments will cause extensive internal damage.

Hydrostatic Shock

Hydrostatic shock is an effect that is often misunderstood and about which a lot of conflicting information exists. The basic principle is that when you try to compress a fluid in a closed container, the container will rupture at its weakest point. Since human beings are largely fluid and encased in skin, it would stand to reason that compressing an area of the body with a bullet would cause an increase in fluid pressure that would damage tissue far from the actual wound.

This does happen to some extent if the right organs are involved and if the bullet is capable of creating a huge amount of pressure. This hydrostatic shock can only be expected from very high-velocity rounds placed in the correct areas. It does not take place in parts of the body which contain gas or air or are spongy. The lungs, throat, and muscles are all highly resistant to hydrostatic shock.

Rifle bullets are capable of causing the greatest degree of hydrostatic shock and also take on added lethality because of the sonic waves and vacuum they create through their high velocity. These multiple effects deserve some special attention since the rifle is presently the most common combat weapon.

Vacuum/Sonic Shock

Studies have found that when the kinetic energy of a bullet is imparted to tissue, the tissue in turn bounces about at a very high speed. This movement of tissue and bullet creates a vacuum into which air rushes. This in turn creates a temporary bubble 30 to 40 times the size of the projectile and a pressure wave of up to 1,500 pounds per square inch lasting for a few milliseconds. (Organs which contain large amounts of fluid or are encased in a tight membrane are more susceptible to such pressure damage.) In order to create such pressure, a bullet must have a velocity between 2,000 and 2,500 fps. Bullets traveling at this speed or faster are classified as high-velocity projectiles and are generally limited to rifles, though a few pistol rounds come close.

The full implications of this transfer of energy are startling. Since liquids can't be easily compressed, the sudden vacuum/pressure changes can cause some organs, especially the liver and brain, to "explode." Rather than injuries being limited to the tissue immediately around the bullet track, damage radiates for some distance from it. Entrance wounds are small but limbs can shatter, muscles can split along their intramuscular planes, blood vessels can burst far from the entrance point, brains can be pulped, and the spinal cord can disintegrate! This effect only happens when a bullet is traveling above the 2,000 fps high-velocity point. Below that speed, most authorities consider the minimum amount of energy required to put a man out of action to be 108 foot-pounds of energy, another useful minimum figure to check out when picking a combat round.

Projectile Tumbling

The ALCLAD studies placed an upper combat-range limit at 300 yards with most combat at below 100 yards. A quick check of the ballistics tables shows that the .308, 30-06, and .223 all have the energy and speed within this range to be effective combat rounds. But the damage done by rifle bullets doesn't end there. Pointed, high-velocity bullets also are unstable and tend to tumble end over end, when hitting a medium denser than air. With military rounds, this is important, because armies are limited to FMJ ammunition by the Geneva conventions. One way to increase the wounding effect of the FMJ bullet is to design a rifle so that the bullet is more apt to tumble. Generally, this is done by decreasing the rifling twist rate. The early M16s had a rate of one twist to fourteen inches of barrel length, which sent the .223 bullet from the muzzle with minimal stability. When the bullet hit, it started to tumble almost immediately. This is why early M16s used in Vietnam gained such a good reputation for taking out an enemy. Like many other things, there is a trade-off with a slow twist rate. In cold weather, .223 accuracy goes down the drain with a 1-in-14 twist. So the twist rate of the M16 was increased and many mourned the decrease in the round's wounding potential.

The new NATO standard for the .223 has demanded an increase to a 1-in-7 twist, which gives stability to the larger bullet being used. Many felt that this fast twist would spell the end of the .223's ability to create massive wounds. New findings by those close to the U.S. Army have revealed that this isn't the case, and that the .223 bullet is actually more deadly now. The increase in twist has placed a greater amount of centrifugal force on the jacket of the bullet. When it strikes, it tends to break apart shortly after entering the skin, creating a multiple projectile-type wound coupled with the high-velocity effect. The pressure wave severs tissue which has been damaged by the bullet fragments and blows them deep into the wound. While soft tissue doesn't do much damage, bone fragments do, resulting in a very dirty and large wound.

The new .223 (5.56mm NATO) round promises to be the most lethal yet fielded for combat. Test results indicate that the bullet will create a wound about ten inches deep with a permanent cavity about four inches in diameter by six inches deep two inches below the surface of the skin, even if it misses hitting a bone. Hits to a bone would create a different wound of equal or greater lethality.

The FMJ .308 Winchester/7.62mm NATO bullet doesn't enjoy the same wounding potential that the .223 round does. The bullet doesn't start to tumble until it has traveled about 7.5 inches into a victim and won't create multiple fragments unless it chances to hit a bone or a metal belt buckle, etc. Thus, while lethal, the FMJ .308 bullet will usually take second place to the new .223/ 5.56 NATO in wounds caused to the extremities.

Many combatants, however, are not limited to FMJ bullets. Such shooters can use bullets with soft or hollow tips designed to expand upon striking flesh. Under such conditions, a single shot with the .308 has a slight edge on the .223. Certainly, the .308 is also capable of penetrating heavier barriers than its .223 counterpart. If you have a .308 rifle, try to take full advantage of its potential by using expanding bullets. (See Appendix E for a comparison of the effectiveness of various types of bullets and calibers.)

PENETRATION

You should try to keep in mind how much penetration a bullet will give you. While it is hard to set a rule of thumb because of differences in ammunition, velocity, barrel twist and length, and bullet placement, it is safe to say that both the .223 and .308 bullets will generally penetrate all soft body armor, glass, and most laminated windows, and be capable of shooting through most frame houses. Here's a chart of the rough figures; they will vary a lot according to bullet shape and design and the type of firearm being used. Bear in mind that these are the maximum effects produced by FMJ bullets. Compacted earth gives about one third or less the protection of sand, and water will increase penetration since it acts as a lubricant.

BULLET PENETRATION
(in inches)

	Compacted Earth	Sand	Concrete
.223 Remington (55 grain)	1.6	5	2
.223 Remington (62 grain, 1-in-7 twist)	4.0	12	6
.308 Winchester	3.5	10	5

The .223 is not always successful on the sides of cars though it will usually make it through the window glass. Some military .223 rounds, as well as those modified as outlined in later chapters, have a steel penetrator that will make it through cars and can cause severe wounds to the occupants. The .308 is capable of good vehicle penetration and will generally reach those inside a car. The .308 is not reliable if you must fire completely through the body of a vehicle, however.

Pistol rounds often vary widely among the various types of bullets available for them. In general, the 9mm Luger and .357 Magnum are quite good at penetrating car doors or other barriers, while the .38 Special and smaller calibers and the .45 ACP lack the power to consistently punch through many such materials. With expanding bullets, a .38 Special with +P loads will penetrate 1-1/2 inches of plywood; a 9mm Luger, 2-1/2 inches; a .357 Magnum, 2-1/4 inches; and a .45 Auto, 1-1/2 inches.

CONCLUSIONS

From the above considerations, we can conclude the following:

- No bullet is totally dependable. There are a lot of if's involved.
- A bullet should be designed to maximize its ability to create a puncture wound. Barrel twist, shot-filled bullets, expanding nosed bullets, etc., will create this effect.
- Small-caliber rifle bullets should have a velocity above 2,000 fps within the 300-yard combat range.
- Military authorities generally agree that at least 108 foot-pounds of energy are needed to knock out an opponent within the useful combat range of rifles, pistols, or shotguns.
- Multiple projectiles are much more effective than single projectiles. The increase in effectiveness is equal to the square of the number of hits.

We'll be looking at these facts later to see how to determine the limitations and capabilities of a combat round you may consider using.

2. A Brief History of Ammunition

While you may not need it in combat, some knowledge of the history of ammunition design and development may enable you to predict upcoming trends and also recognize what has not worked in the past when it is reintroduced in a new, "modern" guise.

Black powder started the proverbial (cannon) ball rolling during the Middle Ages. The Chinese apparently knew how to make black powder but restricted its use to fireworks. Only when Europeans started using it was the destructive potential of gunpowder realized, with its first recorded use in battle taking place in 1346.

Early guns were gigantic crew-served affairs. Only a few small arms were used since they were single-shot and clumsier than even the crossbow to reload. By the 1500s, methods of making strong iron gun-barrels and a reliable cord fuse impregnated with potassium nitrate made it possible for armies to field various types of matchlock rifles. Needless to say, this type of weapon was extremely sensitive to adverse weather conditions and the fighter often relied on the sword or bayonet in combat.

The flintlock was a great step forward. It used a chunk of flint to create a spark which ignited a small powder train leading into the rifle's chamber. The flintlock rifle could be carried ready to fire for long periods without having to constantly light, adjust, and replace a burning fuse precariously mounted on the firearm. Though still moisture-sensitive, the flintlock was a powerful weapon in the hands of both large bodies of troops and individuals.

The flintlock was used for nearly 300 years until its gradual replacement by percussion firearms in the early 1800s. The percussion rifle did away with the flint and the exposed powder trail by fitting an explosive-filled cap over a nipple which led to the chamber of the weapon. When struck by the firearm's hammer, the percussion cap detonated and showered the powder in the rifle's chamber with sparks, quickly igniting it. A number of multiple-shot percussion firearms were also created using revolving barrels and chambers, and were capable of delivering several shots before reloading became necessary.

As metals became stronger, combat firearms became capable of hurling projectiles at higher speeds. With increased speed, bullets became more deadly so that their size could be reduced. Thus, when comparing newer ammunition to various older examples, a gradual reduction in caliber becomes evident.

Warfare demands high rates of firepower. One way of achieving this was to form lines of troops with one row firing en masse while the others loaded. While this was an effective tactic at first, as rifles became more accurate, these lines of troops became good targets. A number of British defeats during the Revolutionary War can be attributed to the inability of the British to adapt their tactics to the improved accuracy of the battle rifles of the time.

Thus, three general trends have marked ammunition development: cartridges have become smaller and also more lethal; the individual soldier has gained greater control of his own actions on the battlefield; and faster reloading systems have been created to give combatants greater firepower.

An early example of these trends was the paper cartridge, which, although it had to be torn apart and loaded manually, freed the soldier from having to measure the powder for each shot and shortened his reloading time. In addition, the spread of rifled barrels and the conical Minie ball made the rifles of the early 1800s much more effective than their earlier counterparts.

Combustible cartridges were also developed during this time. These consisted of a powder charge glued to the bullet and solidified with mucilage or encased in nitrated animal intestines or fish skin. Both types of cartridge allowed the whole assembly to be loaded down the muzzle in one step. This cartridge, coupled with tapes of primer caps, greatly speeded up reloading time.

The next step in faster reloading came with the placing of the primer directly on the powder charge. While this could be done on combustible cartridges, in order to create a more robust cartridge, metal cartridges were used, which held powder, primer, and bullet in one tough, compact package. Brass proved the easiest metal to work with, was inexpensive, and could contain the high pressures created by a fired round.

Primer placement took a number of bizarre turns. The primer was placed in small tube-like extensions on the side of the cartridge and even on the front of the bullet. Eventually, two primer placement methods proved practical: the centerfire cartridge and the rimfire. The rimfire system worked well, but the cartridge did not feed well in many of the new loading systems. The centerfire cartridge, on the other hand, could be made in a rimless design that fed smoothly through a number of different firearm mechanisms. Thus, while the rimfire cartridge enjoyed great success in a wide range of calibers, it has been largely displaced by the centerfire. The .22 caliber rimfire is the only commercial rimfire cartridge at present.

At the end of the 1800s smokeless powder replaced black powder. Smokeless powder allowed the development of lethal high-velocity rounds and also created less of a telltale smoke signature so that a soldier could hide more easily. Smokeless powders were also safer to store and could be produced as cheaply as black powder by a modern industrial country. Black powder's long reign on the battlefield came to an abrupt end.

The future is hard to predict, but current trends are toward smaller, unstable bullets that tumble on impact or shatter to create a number of wound channels; burst modes that allow smaller cartridges to achieve lethality through multiple hits; and lighter metals for cartridges or even caseless ammunition to reduce the weight of a rifle's ammunition. In short, ammunition is becoming ever smaller, lighter, and deadlier.

3. Reloading Equipment

There are a number of reasons to get into reloading. For one thing, do-it-yourself ammunition is usually slightly less expensive for many calibers, so that reloading ammunition may allow you to gain the practice you need to become an expert marksman without spending a small fortune in the process.

However, a number of companies offer ammunition at prices that are quite competitive with what it may cost you to reload your empty brass, especially if you take into account your time and the heavy initial cost of equipment and components. PMC offers ammunition that is equal to most handloads or military ammunition while CCI's nonreloadable Blazer ammunition is available in rounds which, in some calibers, could be used for both plinking and combat. While some inexpensive commercial ammunition isn't as accurate or lethal as what you can reload, it is certainly great for practice. If you are interested primarily in plinking and practicing to gain familiarity with your firearm, you may not need to reload.

The major reason to consider reloading is that it allows you to create loads tailored to your specific needs. With a little equipment and know-how, you can replace or alter bullets, change bullet velocity, and so on, to create ammunition that will increase your chances of winning a gun battle.

Reloading is not as hard as many people think, provided you have a few basic tools and components. Before you run out and start buying equipment, be sure you've read up on everything enough to know what you need. Money goes very quickly whether you're getting the right or wrong equipment.

Basic starter kits for pistol and rifle calibers are offered by Lee and Mequon. These kits consist of a cartridge resizing die, deprimer/primer tools, powder scoop, bullet seater, and other odds and ends. The main advantage of these kits is their cost: they all run under $25, with the Lee kits going for as little as $16 in some calibers. With one of these kits and some powder, bullets, and primers, a person can be reloading in no time.

The catch is that these kits, which use hammer power rather than a press, are slow going, especially if you full-length resize your brass, which you should. You pay for the initial cost savings through the excessive time needed to reload ammunition. These kits do produce ammunition as well as larger presses and pack into a small space. All you need is a good plastic or rawhide mallet and perhaps a powder measure and deburring tool to get started. If you reload only a few rounds at a time or alter small numbers of commercial cartridges and then reload them, then these inexpensive kits make perfect sense. These kits also enable you to reload a number of common calibers in a crisis but are not, of course, practical in battle.

Sooner or later, however, most shooters decide they wish to reload their own ammunition on a large scale. If you are at all interested in reloading, you may end up saving some money by bypassing these smaller units and starting out full scale.

SHOTGUN SHELL RELOADING

Shotgun shell reloading can be carried out quite satisfactorily with a minimum of expense. Shotgun shells are easily reworked since they are principally made of plastic, so good reloading equipment can be quite inexpensive. In 12-gauge, the Lee Load-All Junior is a bargain at under $25. It has a number of stations so that you can reload shells without constantly having to change dies. While the unit is not as tough as more expensive outfits, it does work well and is nearly as fast as presses costing ten to twenty times as much. And it is the only press I know of that can reload 12-gauge slugs without damaging the fingers in the press. All in all, the Load-All Junior is a good buy.

Unfortunately, the Load-All Junior only comes in 12-gauge, but Lee also offers the Load-All, which is actually better than the Junior, costs only slightly more, and boasts a powder holder. Like the Junior, this unit has a number of stations and allows the user to reload with speeds that rival expensive presses. It is capable of handling six- or eight-segment crimps in trap, field, or magnum loads. Both loaders will adjust to 2-3/4- or 3-inch shells and automatically adjust wad pressure. The Load-All comes with a set of replaceable shot and powder bushings to keep it going for quite a while. If you go with the Load-All Junior, take some time to order spare parts so you won't get bogged down should some piece of plastic break. (Though a lot of stress is placed on reloading equipment, the plastic parts are made of long-wearing nylon, which lasts a long time even when the equipment is used extensively.)

I use the Lee Load-All for my shotgun reloading, and I suspect that many people will find that the Lee equipment is all they need. On the other hand, I rely on either an assault rifle or a large-caliber pistol for all my defense needs; those who use the shotgun extensively might be happier with a heftier unit like the Pacific 155 or MEC 600 Jr. For about 90 percent of combat reloaders, though, the Lee Load-All units are ideal.

LARGE-SCALE PISTOL/RIFLE AMMUNITION RELOADING

Pistol and rifle ammunition require a bit more of an outlay to get set up in the reloading business. The brass is harder to rework and tolerances are smaller than on shotgun shells. Most pistol and rifle ammunition reloading operations are built around a quality press. These vary considerably in ease of use, strength, and price.

The Reloading Press

The O press (which gains its name from the O shape of its body) is quite strong but should generally be reserved for swaging operations or larger caliber weapons. It is more awkward to use than other designs since it is necessary to remount each die as it is needed. If you need an O press, Hornady's 00-7, the RCBS Rock Chucker, and the Lyman Orange Crusher are good buys. One of the strongest O presses, designed for case forming and bullet swaging operations, is the C-H Champion; it's expensive, but can handle anything short of heavy caliber swaging operations.

Turret presses allow the user to mount dies and rotate them into place as they are needed. These presses are quite suitable for all pistol and most rifle calibers. (I wouldn't recommend them, though, for any but the smallest of swaging chores.) Probably the best buy is the Lee Turret Press. It has a three-position die holder which can be easily mounted or dismounted. Switching from one caliber to another is quite simple; dies for each caliber can be kept together on a spare die holder. Newer turret presses from Lee also have an auto-indexing option which rotates to the next die automatically. This feature makes for very quick reloading of pistol cartridges and is well worth the cost.

Another excellent turret press is the Lyman T-Mag. The T-Mag has six die-holding holes in its turret and two primer feeds on its base. This allows you to keep the press set up for two different calibers of shells all the time. If you limit your combat needs to two calibers (or three rifle calibers if you use two-die rifle sets), the T-Mag allows you to reload rounds quickly with a minimum of start-up time. The Lyman turret can also be removed so that different die sets can be kept mounted and adjusted. Changing from one turret to another takes only a minute or so.

A useful accessory for your reloading press is an automatic primer feed. This speeds up reloading considerably and makes your ammunition slightly more reliable since you're less apt to get oil on the primers. Primer feeds can be dangerous, however,

Lee Turret Press with auto-index.

The auto-indexing rod on Lee's Turret Press makes reloading pistol rounds quicker and easier.

The turret on the Lyman Turret Press will hold up to six different dies and can be removed and exchanged in just a few moments.

if you allow primer dust to build up in them. This can lead to an explosion that may set off all the primers in the tube. If you're wearing safety goggles—and you should be—all you'll suffer is a damaged tube and a blackened face, but it's a lot better to just keep the dust down and avoid the whole problem.

Progressive Presses

Progressive reloading equipment is nearly automated. A group of dies, powder measures, bullet/brass/primer feeds, etc., are all arranged so that the user just cranks the power lever up and down. In theory, raw materials go in and finished ammunition comes out.

In fact, things aren't quite this simple. If you aren't using carbide dies, your brass has to be lubricated; all the bullets, brass, and primers have to be lined up in their feed chutes; powder has to be placed in its container; and everything has to be carefully adjusted. Most presses also can handle only one caliber of ammunition.

There are operations that can't be carried out on a progressive. It is nearly impossible to check the brass for flaws, clean primer pockets, measure case lengths, or check neck thickness. Because these operations can't be carried out, the ammunition can be considerably less reliable than that produced on the nonprogressive presses. Military brass is nearly impossible to reload since the primer pockets can't be swaged unless a deprimer is used separate from that of the progressive loader.

But the flip side is the speed. Once the unit is set up and the components in place, it is possible to create 300 to 600 loaded rounds per hour. The trade-off for the speed is a more expensive investment and less reliable ammunition. If you're trying to get the maximum number of reloads from your brass, are reloading only small numbers of rounds, switch calibers frequently, or wish to create a large stock of ammunition for combat, then you're better off with either reliable commercial ammunition or ammunition that has been carefully inspected and reloaded on a slower unit. On the other hand, if you need to practice a lot, or are part of a group that shoots large amounts of ammunition, the progressive can make a lot of sense. (Note: You can't legally manufacture ammunition to sell or reload for friends without a special license from the Bureau of Alcohol Tobacco and Firearms. Other legal considerations should be kept in mind; a round that blows up a gun, for example, could drag you into a courtroom nightmare.)

There are two types of progressive loaders. One is the straight-line progressive, on which all the dies and operations are carried out on a bar, allowing the user to see what's going on. This type of progressive loader places a lot of pressure on the front of the press so that heavier bars and parts must be used in order to keep the unit from warping out of alignment. The other style is the rotary progressive press. This is a stronger design, but some of the operations are hidden from the operator behind the press. This means that the machine can be running amok with the operator blissfully unaware of what's happening.

One of the least expensive progressive reloaders is the Lee Progressive 1000, a greatly modified version of the Lee Turret Press. With feeder tubes and automatic powder measure, it is a true progressive, however, and spits out a loaded cartridge with every pull of the lever. One real plus is that the die holder on the Progressive 1000 can be popped off and exchanged as quickly as it can on the standard turret press.

A progressive that costs about twice as much as the Progressive 1000, but is still at the low end of the scale, is the Dillon RL-550. Like the Lee, the Dillon has a removable die holder with four die positions so that changing from one caliber to another is quite simple.

The Lee and Dillon progressive units can handle most shooters' needs. There are better progressive presses available, however, which are easier to use and more ruggedly built. Before purchasing one of the expensive Star Machine Works or Dillon units, which cost up to $3,000, you should determine whether you can tolerate the lower quality ammunition, and whether you need enough ammunition to justify the expense of the machine and the time you spend using it.

RELOADING DIES

The heart of a reloading operation is a set of dies for the caliber being reloaded. Rifle dies usually come in two-die sets while pistol calibers use three-die sets, although four- and five-die pistol sets aren't uncommon. In two-die sets, the first die deprimes and resizes the brass. After priming and adding powder, the second die seats the bullet and may also slightly crimp the case around it.

In pistol sets, the first die resizes and sometimes deprimes the brass, the second bells the mouth of the brass and may deprime it if the first doesn't, and the third seats the bullet and crimps the brass slightly around it. On four-die sets, the third die usually only seats the bullet while the fourth crimps the brass.

Three basic types of resizing dies are available. The two most common are the FL (full-length) resizing die and the SB (small-base) die. The full-length resizer gives the brass dimensions which allow it to be chambered in most firearms. The small-base die resizes the brass to its original unfired size, and the ammunition will chamber reliably even in small chambers or when a chamber is fouled with powder residue or dirt. While small-base dies create ammunition that gives you a slight edge in combat shooting, the ammunition created with full-length dies is generally very reliable and less apt to cause excessive wear on the brass with extended reloading. This is especially true for combat firearms since the manufacturers often cut chambers close to the maximum internal allowances. Because of the extra wear and damage to the brass and the fact that most firearms work well with brass that is full-length resized, use small-base dies only if you're not sure what weapon the ammunition will be fired in or if you have a weapon with a slightly smaller than normal chamber.

The third type of die resizes only the neck of the cartridge. This gives a very slight increase in accuracy (about 0.1 minute of angle–1/10 inch at 100 yards) because the cartridge sits in the same position in the chamber every time it is used, but the ammunition does not chamber reliably in a rifle other than the one it was originally fired in. In semiauto weapons, it is even possible to get a slam fire because of the extra friction created by the oversized brass. The small increase in accuracy and case life does not warrant the use of the

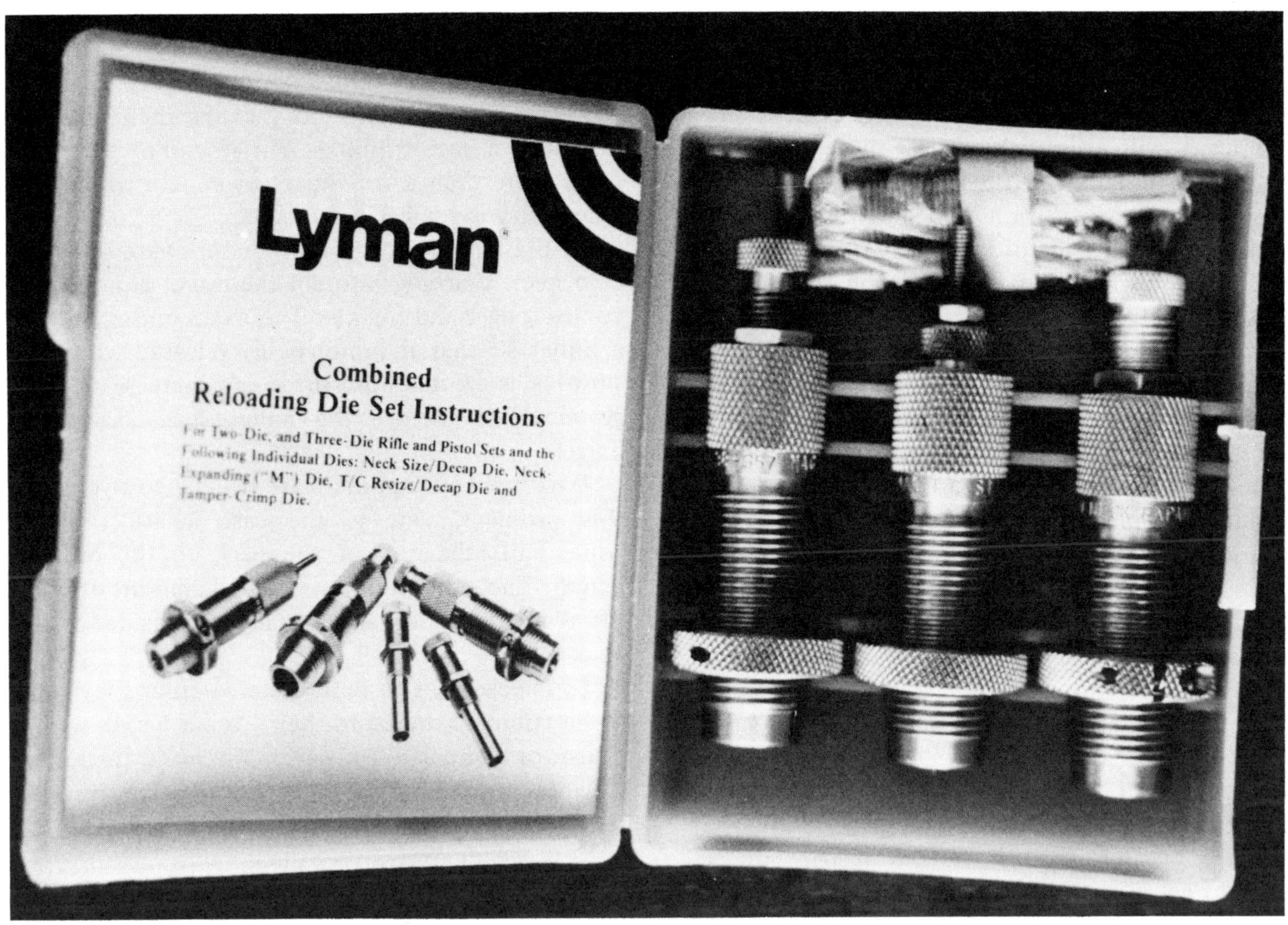

Several companies offer three-die sets for reloading pistol ammunition. Shown here is a Lyman's three-die set for the .357 Magnum.

neck-resizing die in combat ammunition and also isn't even worth the risk of chambering problems with loads designed for target or sniper use. My advice is to avoid neck-resizing dies and go with the full-length dies.

At the time of this writing, carbide dies are normally available in pistol calibers. These aren't actually entirely composed of carbide; in fact, there is only a tungsten-carbide insert in the resizing die while the other dies are made of standard steel. But this insert makes a big difference—it can resize about three times as many cartridges as a steel die before showing signs of wear. The insert is quite strong and doesn't require lubricating the brass before resizing. Chrome-plated dies offer slightly increased resistance to rust formation and are worth considering if you'll be reloading in a humid environment.

There are a number of good dies on the market, though RCBS and Lee seem to have the best reputation. Lee dies are my favorite because the Lee deprimer is heavy enough to deal with the military primer crimp. This allows easy reloading of all types of brass and ensures that the deprimer won't need to be replaced for a long time.

While most dies are good for only one type of ammunition, dies are made that can reload both the .357 Magnum and the .38 Special. Though these cost a bit more, they give you added flexibility for only a few more dollars. The best sets of these dies come from Lyman; though the deprimer is not as tough as Lee's, the extra strength won't be needed since the military primer crimp is rare in this caliber.

SHELL HOLDERS

A shell holder is used to keep the brass in place on the ram as it is pushed in and out of the dies. One shell holder can work with a number of different calibers, so check before buying new ones; you may already have one that will work. Shell holders occasionally come with the die set, but generally they have to be purchased separately.

RESIZING LUBRICANTS

Lubricants are essential for regular steel dies but are not necessary for carbide dies. Not using a lubricant will eventually jam a piece of brass in the die and create a unique paperweight! And—to add to the confusion—too much lubricant will put a dent in the side of the brass.

I use Lee Resizing Lubricant on steel dies. It is water soluble, so it can be easily cleaned off your hands and your resized cases. It dries quickly on brass but maintains its lubricating qualities, so that you can't create a dent by over-lubricating the brass. Best of all, the Lee Resizing Lubricant won't neutralize powder or primers if you should get it inside the brass or the primer pocket. While most lubricants have to be placed on an old ink pad and the shells rolled in it, simply placing some Lee lubricant on a couple of fingers and rubbing it on a case as you twirl it in your other hand is quick and easy to do.

If you run out of Lee lubricant, the best substitutes are anhydrous lanolin or hand soap. Don't try to resize a case without lubricant unless you're using carbide dies; otherwise you'll soon have a cartridge eternally wedded to your resizing die. (If you do get a cartridge jammed in a resizing die, it is possible, with the Lee dies, to remove the deprimer adjustment bolt and hammer the deprimer through the die so that the cartridge comes out. On other dies, it is possible to remove the top and use a bolt or nail of appropriate size.)

When you're finished reloading, it's important to wipe the resizing lubricant off the brass. Failure to do so can increase the rearward pressure created when the cartridge is fired, since the brass won't lock up on the wall of the chamber. This extra backward pressure can fracture a bolt and even cause premature unlocking.

PRIMING TOOLS

Seating depth of the primers can change the bullet's point of impact slightly; it's important to seat the primers at the same depth. Some folks like hand-held priming tools. The theory is that you can feel the primer go into place and have more control over it. I've never found that hand-held units give any more control than the priming set up on most reloading presses. Unless you're reloading without a press, I'd recommend against spending your money on a hand-held tool until you've tried out the press's primer.

CASE TRIMMERS

If you will be reloading brass only once or twice or are altering factory loads, then you will not need a case trimmer. But if you're reloading brass more than three times per case, a trimmer is a necessity for safety reasons.

As brass is fired, it is extruded forward until the case neck that goes around the bullet actually becomes longer and thicker. This extra brass can hold a bullet so that it is not easily released when the powder is ignited; pressures can increase to dangerous levels before the bullet starts down the barrel.

Two types of trimmers are needed to overcome this problem. One is the case length trimmer which cuts the end of the neck to the original length. The other unit cuts a small amount of brass off the outside, or occasionally the inside, of the neck.

To check cases to determine whether they need to be trimmed, measure their overall length with a gauge or calipers. To check on neck thickness, place a bullet into a fired cartridge before you resize it. If it doesn't slip in and out easily, the brass is getting excessively thick.

Forster, Marquart, Dewey, and I.S.W. all offer good outside-the-neck trimmers. The I.S.W. trimmer is one of the quickest to use; it can be mounted in a drill chuck and operated with minimal time and muscle effort.

For case trimming, the Lee Improved Case Trimmer is good for manual operations. It comes with a shell holder which can be chucked in a drill; this is a little slow, however, since the drill has to be stopped and the next cartridge slowly screwed into place. The Forster Power Case Trimmer also allows the use of an electric drill but is also not too fast to use.

A much more satisfactory unit for use by hand or with a power drill is the Brass Barber from Monadnock Machine Company. Though the company recommends that the trimmer be used with the "Case Clutcher," it is possible to hold the brass and carefully push it into the blades, although one slip will mix your fingers with the exposed blade that cuts the brass. The Brass Barber is quick; the brass can be picked up, cut to the correct length, and put down in just a few seconds when a power drill is used.

Top of the line with hand-turned units is the Redding Master Case Trimmer. It has a universal collet that allows any pistol or rifle case to be fastened in place, and the .22 and .30 caliber pilots that come with it cover a wide range of case-trimming needs. The unit also allows the user to clean out the primer pockets. I find that the Brass Barber chucked in a power drill gives the best and quickest case trimming, while the I.S.W. neck reamer in a drill gives a good quick neck thinning. Though other methods are sometimes used, I prefer these two pieces of equipment for trimming tasks.

CALIPERS

A caliper can be used as a case-length gauge to determine which brass needs to be trimmed. The caliper is also very useful for checking the accuracy of swaging equipment, the sizes of bullets, brass, bores, and so on.

Metal calipers are more durable but also expensive; plastic calipers are the type most often purchased by reloaders. A good metal dial caliper is offered by B-Square. The inexpensive plastic calipers sold by RCBS and Lyman are as accurate as their metal cousins but you must be careful not to scratch the plastic jaws when measuring metal objects. A little practice with one of these will allow you to make accurate measurements down to thousandths of an inch.

POWDER FUNNEL

A powder funnel is a necessity if you're not using a powder metering unit. A plastic funnel designed for reloading is the safest to use and allows you to see what the powder is doing in its neck. Get one of the universal funnels designed for reloading all calibers of ammunition.

CHAMFERING TOOL

Many reloaders don't use this tool and the ammunition therefore often gets hung up on the edge of a magazine or the bullets are scarred during reloading, so that accuracy suffers. Chamfering removes the sharp edge inside and outside the cartridge mouth. Hand chamfering tools are offered by a number of companies. Whatever method you use, be sure to chamfer both the inside and outside rim of the cartridge mouth. The extra reliability this gives in the chambering of ammunition in semiauto and automatic weapons is invaluable.

POWDER SCALE

A powder scale gives you a way to double-check what you're doing and is essential if you're working with near maximum loads. One of my favorite scales is the RCBS 5-0-5 which has a magnetically damped arm and allows you to measure powder down to one-tenth of a grain. Maximum capacity is 511 grains, which is more than enough for combat rounds. Lyman, Bonanza, C-H, Pacific, and other companies also offer excellent scales. Whatever model you get, be sure that it has a leveling adjustment. Use the scale on a level surface and keep it covered when not in use to prevent dust buildup.

PRIMER-POCKET CLEANER

If you need reliable combat ammunition, it is essential to clean out the primer pocket so that the ammunition will be sure to fire. While you're cleaning out the pocket, check the primer hole to be sure it is centered and not blocked by debris. Lee makes a good primer-pocket cleaner.

PRIMER-POCKET SWAGER

With GI brass, a crimp is placed around the primer. When the brass is used for reloading, the crimp has to be taken out before new primers can be seated. One way to do this is to use a tool to cut off the extra metal. Take care to keep the blade sharp and not to create a blade chatter mark on the primer pocket. A better tool is the C-H Primer Pocket Swage Die which mounts on a reloading press. Adjustment is a bit tricky and has to be just right or you'll either get a hole that is still too small or rip off the rim of the brass. The die can be mounted in the third hole of the rifle die mounts on the Lee Turret Press, which makes it very convenient to use. The C-H Pocket Swage Dies come in a two-die set to handle both the .223/5.56mm NATO and the .308/7.62mm NATO or .30-06 cartridges.

LOADING TRAYS

A loading tray will help you avoid knocking over brass and keep things orderly so that mistakes are less likely. Trays can be made by drilling holes into a plank of wood, but it is easier and more practical to just purchase an inexpensive plastic tray designed for reloading.

POWDER MEASURES

Some die sets come with a powder scoop that allows you to use a number of powder/bullet combinations. These scoops are accurate and safe as long as you stay within the limits of the reloading data that comes with the die set.

But what if you need to use a smaller or larger bullet and powder combination? This is the time to check the reloading manuals and use a powder measure kit.

Lee and Mequon both offer excellent measuring scoop sets which, used in combination, can create practically any load you need. Provided you're not working with maximum loads, you can work within data tables to get the approximate velocity you need from various powder and bullet combinations which you may not be able to realize with the single scoop included in a reloading die set.

These scoop cups can realize nearly the same accuracy of other metering equipment or even a scale. Care must be taken when using a number of scoops in combination, however, since it is easy to loose track of what you've used or need to use. Over- or under-charging is easy to do if you need to use a number of combinations to get the powder load you want. Use a powder scale to double-check your first trial loads to be sure you're not making a dangerous error somewhere.

Instead of using several scoops, it's a good idea to create a measuring cup that will do your work in one operation. Here's how to do it: measure the powder in the factory scoops and then place them into an old fired pistol cartridge; mark the point the powder comes up to; empty the case and cut it down with a file to the mark; double-check things to be sure the cartridge is cut down accurately; then solder a wire handle to the cartridge. You now have a one-step powder scoop. Be sure to label your homemade scoops if you use more than one; a mix-up could be disastrous.

In some measuring kits, the smaller cups have an open back. It's easy to measure out an excessively large load with these cups, which is very dangerous when it comes time to fire the round you've created. Play it safe and fill in the open backs of any measuring cups with glue.

The fastest powder metering is done with powder measures. These have a powder container on their top and a small lever which drops a metered amount of the powder through the unit. Powder measures do have some drawbacks. They have trouble accurately measuring large-grained powders, and the size of loads thrown by them varies slightly with the amount of powder in the hopper. Avoid using them for maximum loads unless you can check the loads with an accurate set of scales.

C-H, MRC, RCBS, Pacific, Redding, and Saeco all offer good powder measures. Lee's Auto-Disk Powder Measure works well with pistol calibers in the Lee Turret Press; the unit fastens to the expanding die and automatically dumps the powder into the case when it comes up to be expanded.

Whatever type of powder-measuring system you use, be sure to double-check everything, including the type of powder you are using. Extra powder, double loads, or the wrong powder could all lead to disastrous events when firing your firearm.

BRASS POLISHERS

Besides looking good, polished brass chambers a bit more easily than grungy cartridges. Polished brass also extends the life of your reloading dies and is a bit easier to find if you don't have a brass catcher on your semiautomatic firearm.

You can use a liquid brass polisher to get a really handsome finish on brass. Locking a cartridge into a shell holder mounted on an electric drill while you use a Brasso-soaked rag will give a polish worthy of collectors' cartridges. But such a polish is seldom necessary and such a method takes forever if you have more than a small quantity of ammunition to polish.

The best tool is a tumble polisher. Buy a small one and polish brass in several small lots. I've been using a Lyman 600 Turbo-Tumbler to polish brass for some time and have found it perfectly reliable. A number of other brands are also available that are, no doubt, equally good.

Brass tumble polishers can also be used to clean tarnished bullets or small metal gun parts. Don't try to polish steel with brass since the tougher metal will scratch the softer. Don't tumble-polish loaded rounds. Doing so might accidentally fire a cartridge in the polisher. The biggest danger comes from ammunition that makes it through the polisher, however. The motion of the polisher will pulverize the powder inside the cartridge and raise its burning rate to an explosive level. Such ammunition could ignite like a small bomb in your firearm!

Before you start reloading, be sure that all the polishing medium is knocked out of the cartridge and primer pocket.

BRASS CATCHER

Brass catchers have saved me more money in cartridges than I care to think about. Since most firearms used in combat are automatic or semiautomatic, using a brass catcher during practice makes a lot of sense. The best catchers are currently made by E & L Manufacturing for the Mini-14, AR-15/M16, H&K 91/93/94, Uzi, MAC 10/11/11-9, .30 Carbine, KG-9/99, Remington 1100, S&W 1000, and Remington Four/74 among others, and cost $25 each. These catchers make it possible to use a weapon without worrying about lost brass.

Do not use brass catchers in combat unless brass has become impossible to replace.

BULLET PULLERS

Whether you're altering factory ammunition, recycling surplus bullets, or trying to recover components from ammunition that you've reloaded improperly, a bullet puller can be invaluable.

There are two types. One is just a refined collet that tightens around the bullet. You supply the muscle power to pull it out of the cartridge. A better design is available which fits into the reloading press, allowing the user to use its leverage to do the work. Automatic collet pullers are available for presses, too. These are quick to use but often scar the bullet to the point that it isn't usable.

The other type is the inertial puller, which is like a small hammer into which the cartridge fits; pounding gradually causes the bullet to slip out of the brass and it is then caught, along with the powder, inside the hammer. An inertial unit must *not* be used with explosive or incendiary bullets.

Good collet pullers are made by Bonanza, C-H, Forster, and Pacific. Lyman, Quinetics, and RCBS make good kinetic units.

LEAD FURNACES AND BULLET MOLDS

Lead furnaces are used to melt down lead or lead alloys to cast bullets. Because lead bullets produce too much barrel leading when propelled at high enough velocities to be useful in combat, they are good only for plinking and practice and should be avoided with most combat arms. The only exception to this is the lead slug used in shotgun ammunition. (It's also possible to pour your own lead shot, but this process should only be attempted by dedicated shotgunners with a lot of know-how and patience.)

You might use a lead furnace if you cast your own cores for swaging. In that case, you can use one of the furnaces listed below plus the Corbin Core Mould which allows you to create four cores at a time of any practical selected weight. Corbin Core Moulds are available for all calibers so that you can match the core to the jacket you'll be using.

Because of the danger of lead poisoning and the hassle of working with molten lead, I've never used lead bullets or poured my own lead cores for swaging. Most combat reloaders' time and money would be better spent on components rather than an expensive furnace, ladle, molds, etc.

If you do need a furnace, the Lee Production Pot, Lee Bullet Caster, and Lyman Mould Master have pretty good track records. Since a lot of odds and ends are needed in bullet casting, you should also consider buying the Lyman Master Bullet Cast-

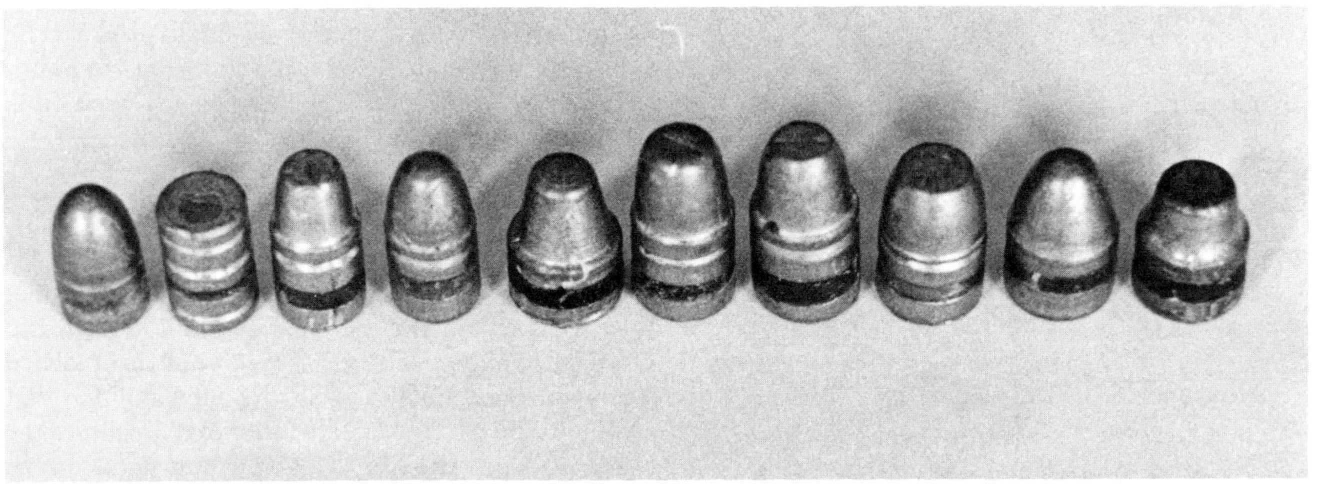

A wide number of bullet shapes and sizes can be cast from the many styles of molds available in the marketplace today. Unfortunately, the lead bullet is better suited to practice than to combat use.

ing Kit which contains the furnace, a mold with handles, bullet lube, top punch, sizing dies, and so on. Lee, Lyman, RCBS, and Saeco all make good bullet molds in a wide range of styles and sizes. Because of the leading problems inherent with the lead bullet, larger bullets are generally used in order to reduce the speed of the bullet and thereby the leading, while maintaining the bullet's power by the extra weight.

A mold has to cast several hundred bullets before it is broken in. Molds also have to be warmed up each time they are used; the easiest way to do this is to cast a dozen or so bullets and then recycle them back into the lead pot. When you've finished with the mold, it's a good idea to leave a bullet in each of its cavities to prevent rust. You should oil molds only when you're going to store them away for a long time. Lead bullets require lubrication in order to cut down on barrel leading. Unfortunately, this lubrication can also get into the powder and neutralize it, especially in hot weather. Thus, it's a good idea to use lubricants which contain little or no beeswax or paraffin. Corbin Dip Lube, Mirro-Lube, and RCBS Rifle Bullet Lube are some of the best for combat use in extreme weather conditions.

Bullet resizing is unnecessary with lead bullets and may even decrease the accuracy of some loads. The slight size difference between bullet and bore is rectified upon firing when the soft lead alloy is swaged to a perfect fit inside the barrel.

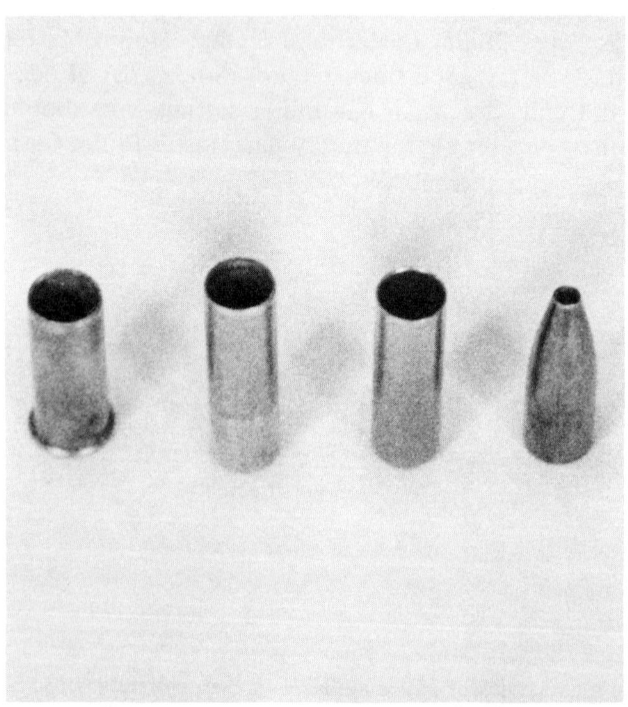

The Corbin Jacket Maker allows the swager to turn a .22LR case (left) to a .224 bullet (right).

SWAGING DIES

Swaging is an art. It doesn't save money or time. It does allow you to make specialized bullets which are not available anywhere else. If you've got a lot of time and are interested in creating special bullets, then you'll love swaging. Dies are expensive, however, and I feel that it's a good idea to really get the feel of reloading before you start purchasing swaging equipment.

The two major swaging die manufacturers are Corbin and C-H. C-H offers pistol dies in .357 diameter (which can be used in the .38 Special and .357 Magnum, and will squeeze fit 9mm barrels), as well as .41, .44, and .45 calibers. The hollow-point die in these sets can also be used to place a hollow point in lead or lead-tipped commercial or cast bullets. The real plus of the C-H dies is that they are threaded for standard reloading presses which develop enough leverage for most pistol-bullet swaging operations.

Corbin offers a wide range of dies for both rifles and pistols as well as special presses for handling the higher pressures needed for rifle bullets of .308 diameter or larger. They, too, offer pistol and small rifle dies in thread adapters for use with standard reloading presses.

In addition to the dies for making swaged bullets, you'll need a good press, preferably a heavy O press, as well as some shop equipment and a lead wire cutter to make the lead bullet cores. Both Corbin and C-H offer very good lead wire/core cutters which allow cutting of very exact lead cores. You will also need to buy lead and copper jackets for the bullets.

Swaging takes a lot of time but can create some very special bullets, as we'll see later in this book. Corbin publishes a number of books which are a big help to beginning swagers. Probably the best is *Discover Swaging,* which can be purchased from Stackpole Books or Corbin.

HOLLOW-POINTER

In addition to creating a hollow-point bullet with a swaging die, there are also tools designed to create a small hole in the nose of bullets or loaded ammunition. Forster sells a hollow-pointer that mounts in its Power Case Trimmer. It allows the user to place either a 1/16- or 1/8-inch hole in the tip of a bullet. A similar device consisting of a die which holds a loaded cartridge in place while a small drill is inserted into the nose of the bullet

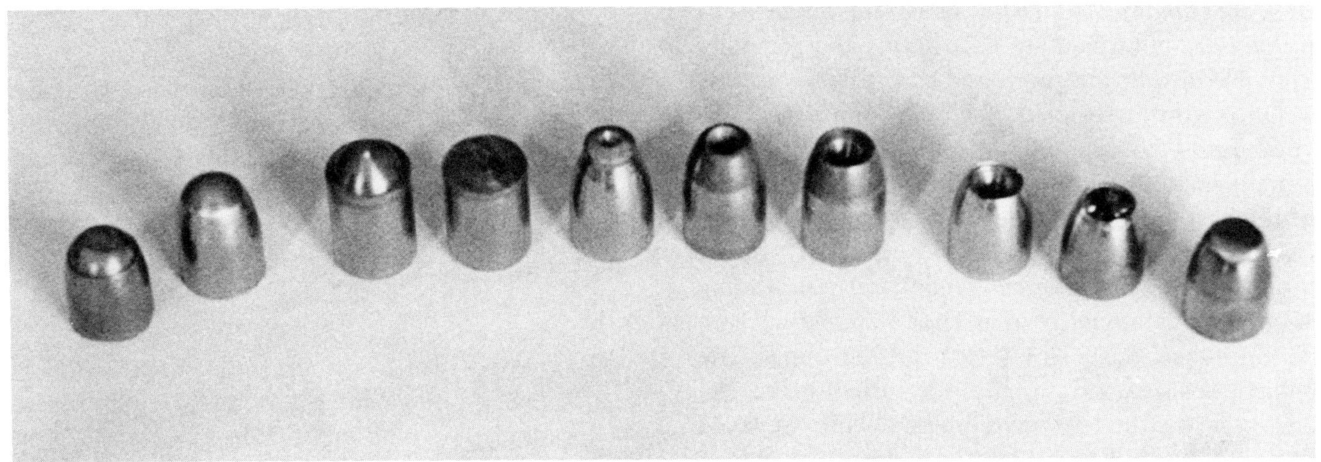

Shown here are just a few of the possible bullet designs one can create through swaging. The author created these for the .357 Magnum by using C-H swaging dies.

Corbin's two-die set for making .224 bullets (left) and the .22 rimfire jacket-making die (right).

is available from Harry Owen.

Care must be taken when hollow-pointing since too deep a hole could create a bullet that would leave its jacket behind in the barrel, causing excessive pressure with a second shot.

RELOADING MANUALS

Many reloading booklets are available free or at nominal cost from companies that sell powder and bullets. While most of these publications only include data on the powder or bullets sold by the manufacturer, they are extremely useful and often list loads that can't be found anywhere else.

A number of books are also put out by companies that do not sell components. These books are good to start with if you're not sure which manufacturer's products to use. Two of my favorites are *Handloader's Digest* and *Metallic Cartridge Reloading,* published by DBI Books and generally available in sporting goods and gun stores.

Once you've selected the best components for your needs, however, it's wise to go with the component manufacturer's reloading data books. Some of the best are the *Hodgdon Data Manual, Hornady Handbook, Nosler's Reloading Manual, Sierra Bullets Reloading Manuals,* and *Speer's Reloading Manual.* You should own at least a couple of them if you're seriously interested in reloading. The ballistic charts and other information make the books a worthwhile investment. While most of these books tend to be a little narrow in that they only show data for their own products, the new *Hodgdon Data Manual No. 25* includes reloading information on powders other than those produced by Hodgdon.

RECORD BOOK

It is essential that you keep accurate data on each load you test or use—if something goes wrong, you'll know what not to do next time. If you come up with a good load, you'll also know how to duplicate it.

Be sure, too, to label any ammunition you store, so that if similar loads develop a problem, you'll be able to locate the stored ammunition and determine what to do with it. Take a little extra time and record what you're doing; later on, you'll be glad you did.

SAFETY GLASSES

You should wear safety goggles when working with reloading equipment. It only takes one small primer explosion or one bit of machinery out of control to permanently injure an eye. Use eye protection when you're shooting, too.

* * * *

It's a good idea to get into reloading slowly. If you're really interested, get a good press, one set of dies, and the components you will need. If you get a Lee die set, you won't need anything else in the way of equipment or manuals since it will include the powder scoop and reloading data for a wide range of loads. Try the works out for a while and see if you're interested in proceeding further. If you find you enjoy making ammunition, you can start branching out into other calibers with more equipment.

Before you start, remember that the equipment and books you purchase are worthless if you don't follow the manufacturer's instructions carefully. Read through it all before you start trying to reload or alter ammunition. You will be working with components that can be used to create explosive devices as well as ammunition—one wrong move and you can quickly damage your equipment, your firearm, and yourself. Don't rush into a disaster of your own making.

4. Reloading Ammunition

Reloading is not very hard, but it is time consuming and, to produce safe and reliable ammunition, requires painstaking attention. Many people who depend on their weapons in life-and-death situations prefer to use factory ammunition in any potentially dangerous situation and reserve their less expensive reloads for practice.

On that precautionary note, let's look at the steps involved in reloading. These are as follows: resizing the case, depriming the case, seating a new primer, adding powder, placing a bullet in the case, seating the bullet, and usually crimping the case slightly to hold it tightly in place.

Reloadable cartridges are made of brass. Brass is far from cheap, and you can reload most economically if you find a good source of inexpensive brass. Military surplus cartridges are good for reloading if you're using the same caliber of ammunition that the Army does. With a copy of *Shotgun News* and a Federal Firearms License (FFL), you can locate someone offering military surplus brass. If you don't have an FFL, you can usually get a gun shop to buy surplus brass for you for a small fee. Another approach is to purchase inexpensive commercial loads for target practice, and then reload them to your own specifications. PMC is probably the front-runner in cut-rate ammunition.

Whatever your source of brass, you should sort cases by type if you're interested in obtaining maximum accuracy. The improvement gained from sorting pistol ammunition is most noticeable, with grouping and accuracy improving by about half an inch at 25 yards. With rifle ammunition, accuracy improves by about 3/10 inch at 100 yards (though this varies with the caliber and rifle). Sorting the cases is not necessary if you will not be zeroing your weapon when you change from one ammunition lot to another. If you're working with maximum loads, you *must* sort cases and check each lot to be sure excessive pressures aren't being created. You may find that certain cartridges—especially military brass—have less room for powder and may create dangerously high pressures.

There are several ways to sort brass. One way is by the head stamp on each cartridge. This will indicate the manufacturer and, with military brass, date of production. While you're sorting cases, inspect them for cracks, major dents, split necks, loose primers (which means the primer pockets are too large), off-center primer holes, stretch marks along the case head, deep scratches which may cause a case to split upon firing, or excessive corrosion. Discard damaged cases at once.

The areas to check for splits in brass are along the neck and upper body where a break in the metal will run vertically. Separations will occur around the circumference of the case, usually about 3/8 to 1/2 inch from its head. Potential separations will often appear as a telltale ring of lighter colored metal around the head; if you see this, discard the cartridge.

With cartridges that have been used a dozen or more times, use a bent piece of wire to feel the inside of the case. The L-shaped wire should be

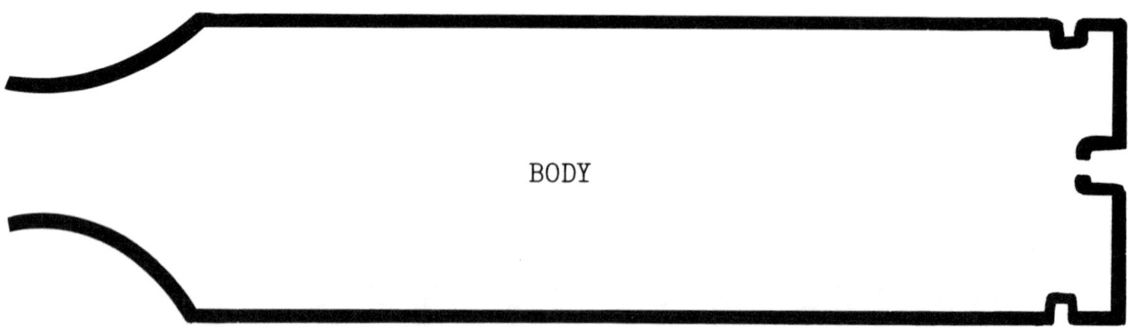

Brass cartridge case

longer than the case with the shorter end about the size of the cartridge mouth, and ending in a sharp point. By sticking the point into the cartridge and pulling it up the inside, it is often possible to discover a developing separation crack.

If there are a number of different head stamps in a cartridge lot, you may want to use a different sorting method. Since the outward dimension of cartridges is relatively constant, changes in interior space are created by excess brass, so that cases with smaller inside dimensions can be detected by weighing them. While this takes time, it will allow you to locate brass that should not be used for maximum loads and to mix case lots while maintaning a high level of accuracy.

You should also sort pistol or straight-walled rifle cases according to whether they have a cannelure, a small, "treaded" indentation around the case body which keeps the bullet from slipping back into the cartridge during chambering. Cannelures can cause the brass to stretch somewhat during resizing, so that it will need to be trimmed more often than cases without them.

If you're retrieving brass after shooting, always sort it according to how many times it has been fired. After the third firing, you should inspect cases for cracks and check the case lengths and necks for excessive brass. When several cases in a group start to crack or show other signs of deterioration, discard the whole lot or anneal them (see Chapter 7). If your brass has been on the ground, it should be cleaned to remove grit which might damage reloading dies. Tumble polishing is best, but if this is not practical, at least wipe the cases off with a clean rag soaked in soapy water or alcohol.

Brass that isn't fired with maximum loads lasts for quite a while if you give it a little care; you should get five to a dozen or more reloads from each cartridge before you have any problems with it. At the same time, reloads should be relegated to practice use after several cyclings since you don't want to risk a malfunction during combat. Don't stake your life on brass that's been used that much.

If the cases have been fired three or more times, then it's time to check mouth thickness. A bullet can be used to check for tightness in the neck before the cartridge is resized; if you can't easily push the bullet in and remove it, it's too tight, and should be reamed. When reaming, be careful not to remove excessive amounts of metal; just take off enough to eliminate excessive tightness. You may want to mark reamed cases with a felt-tip marker and check them at the next reloading to see if you're removing enough brass. It is also a good idea to chamfer the case mouth to ensure easy bullet seating. Chamfer the outside rim of the mouth, too, on rounds used in semi- or full-auto weapons for easier chambering.

The next step is to lubricate and resize the cases. To lubricate the cases, some reloaders put lubricant on their fingers and twirl cases by hand; others roll cartridges over lubricant-soaked ink pads; some put a rag soaked in lubricant into a plastic bag with the cartridges and shake everything up until the cases all have a light coating on them. Pick a way that works for you; just don't omit this step unless you use carbide dies. Remember that too much lubricant will create dents in the cartridge when it is resized, and oil will deactivate both the primer and powder—so don't get oil-based lubricant into the wrong places.

During resizing, depriming is also generally carried out. When the primer has been removed, inspect the primer pocket and clean it out if it contains soot or burnt powder flakes. If you're reloading military brass for the first time, you'll need to swage or otherwise remove the primer crimp at this point.

After resizing, check the case lengths of the cartridges if they've been fired a number of times. Get out the calipers or a cartridge gauge and check for maximum length (you can find the maximum and minimum lengths for your cartridge in your reloading manual). If the length is beyond the maximum, the cartridge should be trimmed and the new edge rechamfered. While bottle-necked cartridges can be trimmed to their minimum lengths during a trim operation, straight-walled cases, which are headspaced on the neck, should be trimmed to a cautious compromise between the maximum and minimum lengths. When resizing and trimming are finished, clean the case of any lubricant that may be left on it.

If the cartridge has been used a number of times, you may wish to clean out the neck to prevent powder residue from making the bullet hard to seat. A small bore cleaning brush of the same type used to clean the firearm is ideal for this step. If you place it in a drill chuck, the cleaning can be done quickly, though you may prefer to use an old screwdriver handle or the like. Do not lubricate the inside of the neck with oil-based lubricants, which may make the bullet seat more easily, but can also ruin the powder or cause the bullet to slip back when it's chambered. Either can be a disaster in combat.

At this point, the primer is seated. Don't have more than one type of primer out in the reloading area at a time; it's easy to start using the wrong one by mistake. A lot of other mistakes can take place when the primer is being seated, especially with presses that have a lot of leverage or in progressive units. Be sure the primer is seated fully (but not over-seated) and is not upside-down or sideways. Also, try to not touch the primers at all with your fingers; oil deactivates them. Use tweezers to load primers into a priming tool one at a time.

Powder is placed in the cartridge next. If your reloading system requires that you remove the cartridge from the press for this step, use a procedure that will allow you to place the bullet over the charged cartridge immediately or otherwise prevent a double charge (or omission of a charge). If you place primed brass in a reloading tray, keep all the shells empty and charge them all with powder at one time to avoid skipping or doubling up on any cases. Whenever possible, visually inspect the inside of the cases or use a matchstick to check the powder height in each case. You can't be too careful with powder charges; double-check everything.

Be sure you also double-check the powder loads with a powder scale, if you have one, about every ten charges. If you use a powder scoop, you may wish to level the cup off with the edge of a card for maximum consistency. Don't have more than one type of powder out in the reloading area at a time; it's easy to start using the wrong one. Also, check the reloading data against the powder, bullet weight, and cartridge type several times; it's easy to get onto the wrong page or wrong column, etc., and load up some real doozies.

Powders made by different manufacturers sometimes have the same number tacked onto them. They *are not* the same powders, however. Don't use a Hodgdon (or whatever) powder with data for DuPont powder having the same numerical suffix—you're apt to blow something up.

Another mistake beginners often make is trying to load the maximum powder charge shown on the charts. Even done correctly, this causes excessive wear on a firearm fed a steady diet of the ammunition. If a mistake is made or components create higher than expected pressures, it can get pretty dangerous when the published maximum is approached. When you add in the possibility of human error, case dimension differences, inaccurate equipment, and so on, the chances of getting too hot a load become very great.

Dangerously hot loads are also hard to "read." While a number of manuals advise you to check the primer for flattening or to look for brass that has melted into the extractor area, it isn't all that easy to recognize what you're looking at without experience. With semiauto actions, you may not be able to find the first critical empty case that might warn you of approaching trouble. Autoloaders sometimes do give you an extra clue by creating a stovepipe jam with a hot round since maximum loads often cause extraction problems.

To top it off, maximum loads will greatly reduce barrel life and often cause excessive parts breakage. My advice is to forget maximum loads and stay within more conservative limits to save wear and tear on your weapon and give you one less thing to worry about. If your firearm won't do the job with anything but a hot load, you need a larger-caliber weapon.

The other extreme is just as bad. We'll be looking at squib loads later, but the basic idea is not to use the minimum loads given in most reloading manuals. At best, they may not cycle semiauto weapons and at worst they may lodge a bullet in the barrel. Like a lot of other things, the middle of the road is the safest route to take with powder charges.

While most powder charges don't fill the cartridge, some come pretty close so that the next step, seating a bullet, can make things pretty tight. One way to get around this is to tap the charged case a bit so that the powder settles before seating the bullet. If you use a funnel, it is also possible to place a long pipe or funnel extension on it so that the powder drops a way before reaching the cartridge and packs itself into a smaller space.

For maximum reliability, *always inspect the powder level* in the cases before seating a bullet. A lot of problems with reloaded ammunition are caused by either double loads of powder or none at all. Either one is disastrous in the best of times and in combat can be fatal for the user. If you can't inspect the powder load, weigh the finished shells and lay aside any that are too heavy or too light.

With the shotgun shell you'll also add some sort of wad between the powder and the shot load at this point, and the cartridge will be crimped over the shot. But with rifle and pistol ammunition, you are almost through. As with primers and powder, don't have any bullets other than those you're reloading out in the reloading area. It's easy to seat the wrong ones.

Don't seat the bullet too far into the cartridge. This can create extra high pressures and can even be dangerous with small sized, high-pressure rounds like the 9mm Luger.

Bullet seating depth depends on bullet shape and the proper length needed for the cartridge to clear the magazine in semiauto action. A rounded bullet will have to be seated more deeply than its pointed counterpart if it is to chamber without getting caught in the rifling of the weapon. After seating bullets in the first few rounds, check them for maximum and minimum recommended overall length and then try cycling them through the action of your firearm, remembering that they could accidentally go off. It is therefore imperative that you keep the weapon pointed in a safe direction. Generally, the farther out a bullet is seated, the more accurate it will be. Accuracy varies greatly from one firearm to the next and bullet seating must suit the firearm for optimum accuracy.

After the bullet is seated, a crimp is generally added to the rim of the case mouth to hold the bullet in place. With bottle-necked or rimmed ammunition, this crimp can be quite pronounced; with cases that head-space on the rim (like the .45 Auto and 9mm Luger), it must be very light. Crimping is especially important with cartridges that will be used in autoloading rifles or magnum revolvers. A poorly crimped round may cause a rifle bullet to be pushed back into the cartridge when it is chambered or allow soft-point bullets to shake loose in a magazine and damage their points, which can result in failure to chamber or excessive pressure upon firing. With pistols like the .357 Magnum and .44 Magnum, the weapon's recoil can cause poorly crimped bullets to drift forward. This can lock up a cylinder and put a revolver out of commission after the first shot.

Cannelures on the bullets are nearly essential for combat rifle rounds and good on all bullets. They allow the crimp to lock the bullet more firmly just an extra bit and provide added reliability. It is possible to add the groove to bullets without cannelures. Both Corbin and C-H offer tools which work with all calibers.

The down-side of cannelures is that they dictate the bullet's seating depth, although the depth can be adjusted somewhat by working with the trim lengths of the cartridges. Trimming to the minimum allows lower seating in the firearm's chamber while the maximum moves the bullet forward, closer to the rifling. Just be sure the cartridges are all trimmed to the same length and that the maximum length still allows cartridges to feed easily through the magazine. Don't try this with cartridges that head-space on the cartridge mouth; with these, try to place the cannelure on the brass itself at the same point as the base of the bullet. Minimum flaring of the mouth of this type of brass also helps.

If you notice small pieces of brass or copper jacket left near the cartridge mouth during seating, then the mouth needs to be opened more fully beforehand. With most pistol die sets, it is possible to adjust the mouth flaring; with rifle ammunition, using boat-tailed bullets will help prevent this problem as will deeper chamfering of the inside of the mouth.

When the cartridge is loaded, inspect it carefully to be sure the primer and bullet are fully seated and the cartridge isn't cracked or dented, and give the bullet a push or pull to be sure it is tightly in place. If the bullet isn't tight and can't be tightened with a second run through the crimping/seating die, you've probably stretched the case mouth during chamfering or trimming, or reamed out too much metal, leaving the neck too thin. Removing the bullet and powder and running the cartridge through the resizing die may rectify the problem. If the sizing die also has a deprimer, the primer can be removed and reseated and will usually work.

It's a good idea to give the loaded cartridge a final careful cleaning to be sure all lubricants and dirt have been removed. It is important that the cases be clean of dirt, powder, and oil if they are to function at their peak. A rag soaked with alcohol, lighter fluid, paint solvent, or the like works well.

If you have any rounds which you think may not be loaded properly, or which appear to have a problem, dismantle them with a bullet puller and redo or discard them. If you have to discard loaded rounds, try to either pull the bullet out or push it into the cartridge and then deep six the whole thing in oil or water. Placing a cartridge in water with the bullet in place will not deactivate the shell for some time—often for several weeks if you've done a good job on it.

Which brings up another point. A lot of people who plan on using ammunition under adverse conditions or who will be storing it for extended periods try to seal each cartridge to prevent moisture damage. This probably isn't necessary unless the ammunition will be buried for decades in a moist environment, in which case the brass will be so corroded that it won't be serviceable anyway. However, if you're really paranoid about this, you can use shellac or fingernail polish to seal the mouth-to-bullet interface and the primer pocket, but you're probably wasting your time.

Before you fire your ammunition or store it, take some time to log what you've loaded, what powder was used, etc. If the ammunition won't be fired for a while, place a note with it indicating what the powder and so forth is so you'll know what you have when you break into the package six years from now. It's also useful to mark brass to distinguish special loads, squib loads, tracers, etc. A temporary measure is to use a Magic Marker on the base, neck, or side of the brass. More permanent markers include shellac or fingernail polish. Polish works well since it comes in a wide range of colors. You can color the tip of the bullet, or kill two birds with one dab of fingernail polish by color-coding around the primers or bullet/neck area, which tightly seals the bullet against moisture. With the wide range of colors available, you should have no problem coding all the types of ammunition you may create.

PRIMERS

A number of primers are available for firearms. Most shooters will find little difference between the various types, but once you find a bullet/powder combination that you're happy with, you should try experimenting with different primers. Chances are, one brand or another will give you slightly improved performance with your pet load.

There are large and small primers for both rifles and handguns. Don't ever try to substitute pistol primers for rifle primers; you're apt to get slam fires or pierced primers. Rifle primers will generally fail to ignite in pistol ammunition.

The various types of primers have slightly different effects during the ignition of powder in a cartridge. Some primers are designed to ignite powder quickly while burning longer; these are the magnum primers. Magnum primers *cannot* be freely substituted for standard primers as they can raise chamber pressures and may even create dangerous pressure levels with near-maximum loads. My experience indicates that magnum primers are usually not needed in normal combat calibers. Whatever primer you use, if you change primers, drop back on the powder loads if you're working near maximum. Failure to do so may be dangerous.

Old-style primers have rounded faces. These have all but vanished from U.S. ammunition. New primers are flat. The Boxer primer is pretty much standard in the United States since it is easily removed for reloading. Europeans favor the Berdan primer, which requires a special tool for removal before a new primer can be seated. Because of the need for the hard-to-obtain primers and tool, most reloaders are better off avoiding Berdan-primed ammunition or treating the brass as nonrecyclable. (The Berdan primer is also used on aluminum cases; these cases should never be reloaded as they become weak and brittle after one firing.)

Proper primer-seating depth is just below the edge of the cartridge. Seating primers too deeply will often cause misfires since the firing pin can't reach them. A primer that sticks out may cause a slam fire.

Live primers can be removed if a press is used

with a gentle technique. This may not work if the primers have been crimped into place. Trying to remove a live primer with an inexpensive decapper and a hammer will almost always cause it to go off.

POWDERS

A wide range of powders is available to the reloader. Generally, the faster powders are better in combat since they will guarantee high velocities with the shorter barrels found on most military firearms as well as minimizing flash signature. They will also create less fouling with gas-operated actions. (See Table I.)

TABLE I: COMMON RIFLE AND PISTOL POWDERS

(in order of fastest to slowest burning rate)

1. Norma R-1	28. Norma 200
2. Hercules Bullseye	29. DuPont IMR 4198
3. Winchester 231	30. Hodgdon H4198
4. Hercules Red Dot	31. Hercules Reloader 7
5. DuPont Hi-Skor 700X	32. DuPont IMR 3031
6. Hercules Green Dot	33. Norma 201
7. DuPont SR 4756	34. Hodgdon H322
8. Hercules Unique	35. Winchester 748
9. DuPont SR 7625	36. Hodgdon BL-C2
10. Alcan AL-5	37. Hodgdon H335
11. Hodgdon HS-5	38. DuPont IMR 4895
12. Alcan AL-7	39. Hodgdon H4895
13. Hodgdon HS-6	40. DuPont IMR 4064
14. Winchester 540	41. Norma 202
15. Hercules Herco	42. DuPont IMR 4320
16. Hercules Blue Dot	43. Hodgdon H380
17. Alcan AL-8	44. Winchester 760
18. Hodgdon HS-7	45. Hodgdon H414
19. Winchester 630	46. Hodgdon H205
20. Hercules 2400	47. DuPont IMR 4350
21. Norma R123	48. Norma 204
22. Hodgdon H110	49. DuPont IMR 4831
23. Winchester 296	50. Hodgdon H450
24. DuPont SR 4759	51. Hodgdon H4831
25. DuPont IMR 4227	52. Winchester 785
26. Hodgdon H4227	53. Norma MRP
27. Winchester 680	54. Hodgdon H870

As mentioned previously, powders made by different manufacturers with identical numbers are not the same. They may be similar, but they are different enough to get you into trouble if you casually substitute them without regard to the differences.

Another mistake can occur when someone reads that some commercial ammunition is loaded with X grains of ZZZ powder. Since the company offers ZZZ powder, the reloader buys a case of it and starts loading his own cartridges using the same brass and bullets. The problem is that companies modify the commercial powder offered to reloaders to conform to powder of the same type number from previous years; otherwise, old reloading data might be dangerous to use. With powder used "in house," however, the company doesn't have to match specifications; they can just reduce or increase the powder load to get the proper chamber pressure and bullet velocity. In effect, the ZZZ powder they use in their own ammunition may be completely different from the ZZZ powder they sell to reloaders. The results can be as dangerous as if the wrong data for the powder were used for reloading. So don't try to second-guess the manufacturers. The same goes for pulling bullets from commercial or military ammunition and deciding that you know what powder they're using. It may look like some other powder, but chances are it's quite different. When measuring powder, take the time to get it right and never trust your memory as to what the correct powder or powder loads are.

Powder will gradually lose its power over time. This can be minimized by keeping the powder cool, dry, and stored in its original airtight container. Don't place powder in heavy containers, or put cases of powder in a heavy metal cabinet. Powder containers are designed to rupture and allow the powder to burn if exposed to a fire, but if the powder is sealed in a heavy container, the effects can be explosive.

BULLETS

Bullets come in a wide variety of shapes and nose configurations. Currently, work is going on in the development of sabot rounds, bullets that will veer off impact, and nestled bullets which will create multiple impacts on a target. Each round promises greatly enhanced lethality over current designs.

Probably the most lethal bullets now on the market are the "Safety Slugs," metal shells filled with small shot. Though these aren't available to reloaders, they can be created by swaging. Runners-up in the lethality race are hollow-point bullets, with soft-point bullets (bullets having an exposed portion of the inner lead on their nose) coming in a close third. For those who are limited to FMJ ammunition, changes in bullet design and barrel twists allow the round to tumble on impact or break apart under centrifugal force when hitting flesh. All in all, modern bullet design has made most calibers much more potent than they were a few decades ago.

There are four basic shapes of current commercial bullets: the flat-nose, round-nose, pointed-spitzer, and boat-tail bullet. Variations on these four types are created with jackets, nose types (hollow point, soft point, or FMJ), base design (flat, hollow, or boat-tail) and type of material used inside the bullet, which can vary from common lead to brass, steel, mercury, grease, or combinations of these.

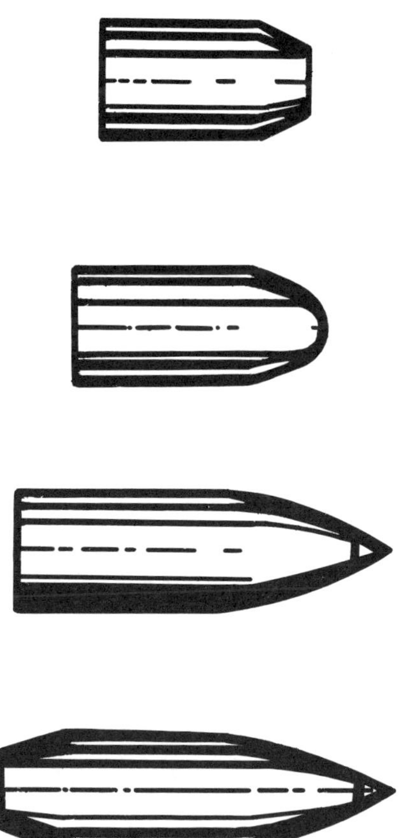

Basic bullet types shown here, from top to bottom, are the flat nose; FMJ; flat-base, pointed spitzer; and boat-tailed, pointed spitzer.

The least effective combat pistol bullets are the round-nose types and for rifles it is the FMJ pointed bullet. Hollow-points seem to be the best commonly available compromise and tend to be slightly more accurate than other types. Since autoloaders can greatly limit the type of bullet that will feed through their actions, it's a good idea to have a gunsmith rework such firearms to function with various configurations. Most automatic pistols can be made to chamber an empty cartridge and will digest almost as wide a range of bullets as a revolver will.

Thin-jacketed bullets designed for varmint shooting are unsuitable for combat rifles. These bullets explode if they hit light obstacles and don't offer enough penetration on impact. They often have an "SX" (Super eXplosive), "Varmint," or other designation.

One good way to determine a commercial bullet's combat suitability is to check the literature the company puts out about it. Bullets are designed for specific purposes, and finding a bullet designed for hunting deer-sized animals or for self-defense will give you a head start in creating a good combat round. Varmint bullets might be of use with very short-barreled combat rifles, which often have a bullet velocity too low to allow standard rifle bullets to expand on impact, while varmint bullets may have ideal expansion characteristics at the lower speed.

Lead bullets are generally unsuitable for modern combat weapons. Because the wounding potential of modern small arms is created by moving a lower mass projectile at a very high speed, the fouling created by lead bullets at such speeds makes them very unsatisfactory. As we'll see later, however, lead bullets used in light squib loads have some limited use in silenced weapons, for quiet practice, or for teaching beginners how to operate a weapon.

Boat-tail bullets do not seem to cut down on barrel life while they do add to long-range accuracy. They are generally the best bullets to use in rifles, especially sniper weapons. However, some rifles are more accurate with flat-based bullets despite the edge that boat-tail bullets normally have. If accuracy is of prime importance, try out different types and weights of bullets to find the one that works best in your particular firearm. Different bullets of the same caliber and weight but with slightly different shapes do not create the same internal pressures with a given powder charge. Thus, you cannot use bullet data without also considering bullet shape or substitute a heavier bullet using a lighter bullet's reloading data. If you do so, you may create dangerously high chamber pressures.

Usually, the shorter a bullet is for its weight, the more accurate it will be. The trade-off is that its velocity will not be maintained as well over

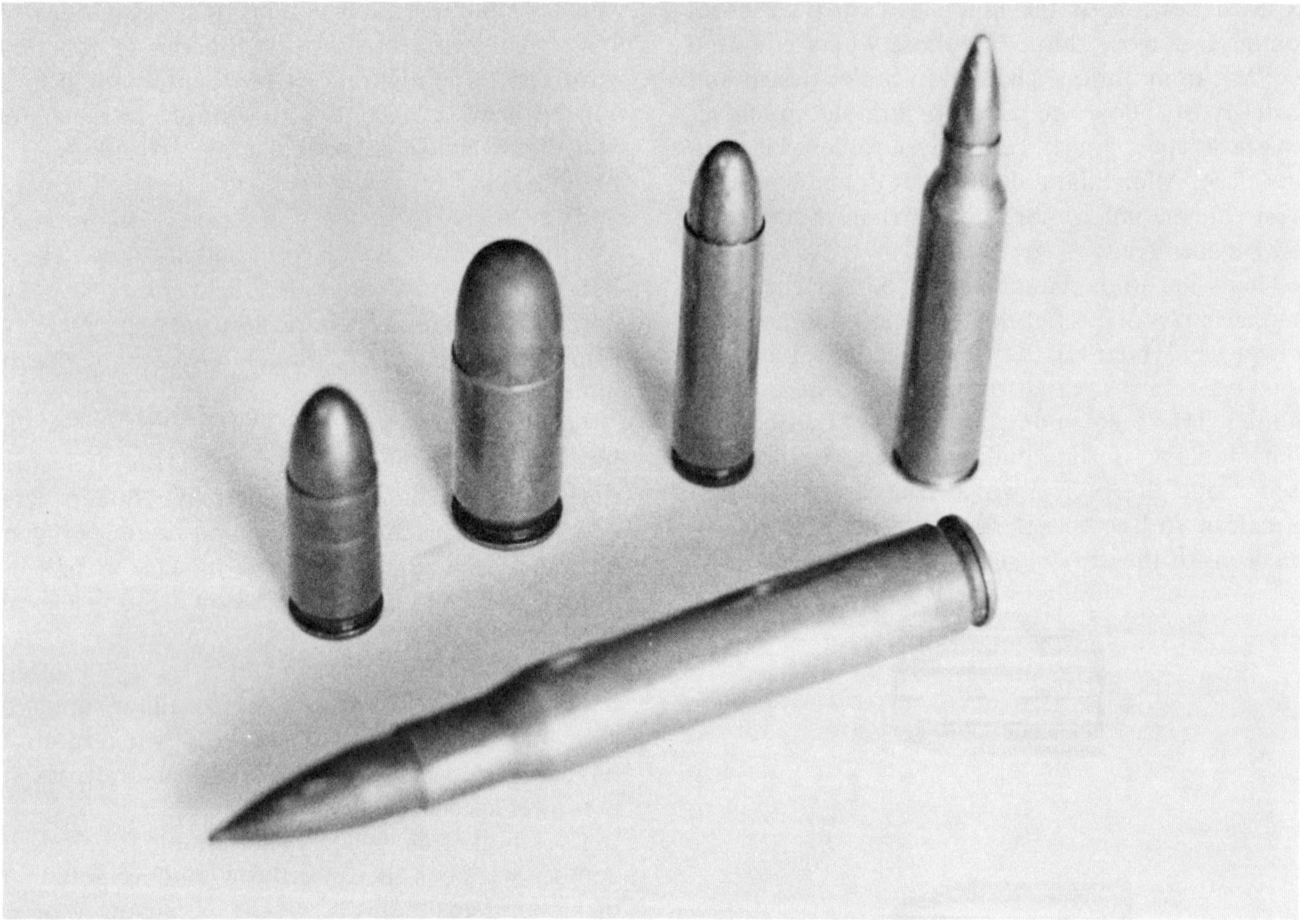

No matter what the caliber, the FMJ bullet is never as effective as its expanding counterpart.

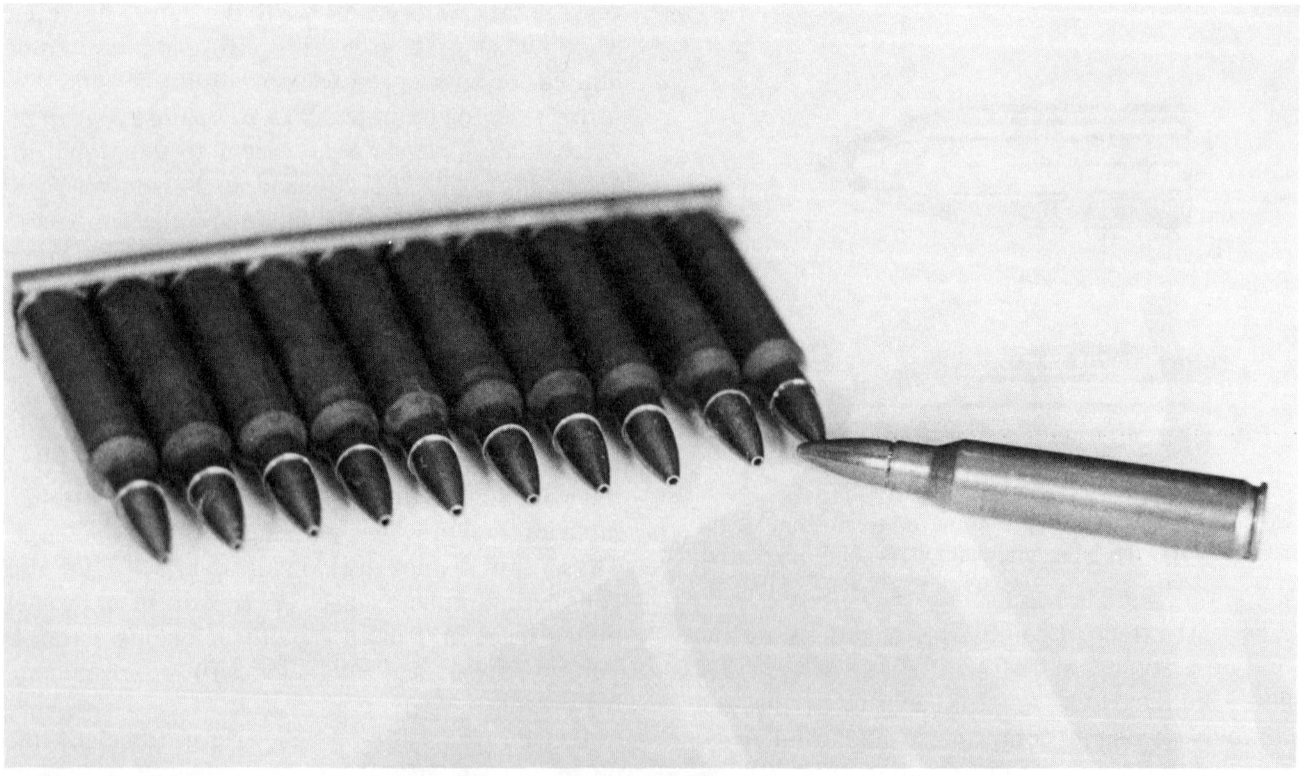

Each hollow-point bullet on this clip of ammunition is vastly superior to the single FMJ load shown by itself at right.

extreme ranges as its thinner counterpart. A good example of this comes up when a short 70-grain hunting bullet is fired from the old AR-15; with a 1-in-12 twist, the bullet is quite accurate. If the new, slightly lighter 5.56mm NATO round is used with this twist, accuracy goes out the window. A faster twist isn't necessary for a larger bullet if its length isn't greatly increased and if you don't mind a loss in long-range power.

Two terms you may encounter are "ogive" and "ballistic coefficient." A bullet's *ogive* is the amount of curve of the bullet's front end compared to its diameter. A 6-caliber ogive means that the curve of the nose, if extended into a circle, would be 6 times the diameter of the bullet. Generally, ogive is not as important to the reloader as is the ballistic coefficient. The *ballistic coefficient* is used as an index to show how well a bullet travels through the atmosphere. The larger the ballistic coefficient number, the better a bullet will maintain its velocity, wind resistance, and accuracy. One catch is that the ballistic coefficient changes slightly with changes in the bullet's speed. So if you're using a bullet designed to travel at a different speed, like a small bullet designed for a different type of cartridge, its ballistic coefficient may go down with a higher speed.

Finally, whatever bullet you choose, be sure that it really does what it is supposed to do when it reaches the target. Occasionally, you may find that a bullet with lots of speed and an ideal nose configuration actually fails to open up, greatly degrading its ability to wound. Be sure, therefore, to test combat rounds in the manner outlined elsewhere in this book. Your firearm, powder, or reloading skills may make what is touted as the best thing since smokeless powder into little more than a chancy shot. Know what your ammunition can do; *test it out.*

LEAD BULLETS

Lead bullets are useful mainly for target practice. They can't sustain the high velocities needed in most combat rounds. Alloying lead with tin or other metals creates a bullet which can travel at higher speeds without coating a bore with lead, but then there's another problem: the bullets don't expand. If, however, you had to make do in combat with lead bullets soft enough to expand, it is possible to trade the high velocity/small bullet combination for a low velocity/larger bullet combination. A bit of extra velocity can be pushed out of lead bullets by putting copper gas checks on their bases or using harder alloys. But overall, less energy can be obtained with lead bullets and, thus, the lead bullet's combat potential is generally inferior to that of its small, jacketed counterpart.

Several companies offer lead bullets to the reloader. These are relatively cheap and offer many of the benefits of inexpensive practice without the hassle of casting. Probably the best known company for these bullets is Alberts, though there are several other good companies as well. Some savings can be realized over time by casting your own bullets, but this saving is not as great as one might hope since the equipment investment can be quite high and many sources of free or cheap lead have dried up.

One of the main precautions to take in casting lead bullets is to have adequate ventilation. Lead poisoning is a gradual thing; people may have mild cases without realizing it. It lasts a long time, since the human body stores lead in bones and nerves, and can literally wreck the nervous system. Some lead is vaporized when it is melted. This vapor can be inhaled and deposited in your body, or the gas may cool, solidify, and coat objects in your home with a thin layer of poisonous dust which is easily ingested or inhaled. The bottom line is: *either cast bullets outdoors or use adequate ventilation.* Household ceiling and kitchen exhaust fans are *not* adequate ventilation.

The worst enemy of bullet molds is rust. When you're storing molds for a short time, leave bullets in them. If you'll be storing molds for a long time, soak them thoroughly in oil. For extended storage, you might even submerge the molds in an oil bath. Newer molds get around this problem by being made of aluminum. These molds are a bit easier to damage, but the elimination of the rust problem makes up for the extra care needed to avoid banging them up. I'd suggest you get an aluminum mold if you can find one in the bullet style you want. Lee Precision and Northeast Industrial both make aluminum molds.

Don't use metal objects to scrape off lead that sticks to a mold. They can scratch the mold or throw it out of alignment. Either use wooden sticks to scrape off lead, heat the mold up so that the lead melts off, or pour molten lead on the spot and carefully wipe off the excess with a large rag.

When you cut off the sprue (the top part of the lead which extends over the mold) with the sprue cutter, don't use a mighty stroke. Heavy blows will eventually bend the sprue cutter or throw the mold out of alignment. Use a wooden or plastic mallet and a number of gentle taps. This takes longer but it beats buying a new mold.

When the lead bullets are hard enough to be released from the mold, they are still not as hard as they will be when they cool completely. If you just drop them onto the workbench, they'll be slightly deformed by the fall. Therefore, gently drop the bullets onto a heavy cloth in order to keep them in good shape until they are cooled and hard enough to be moved.

If you've oiled the mold, it's possible to place several loads of molten lead in it to burn out the oil, then recycle these bullets into the melting pot until the mold is cleared of oil so that the bullets being cast are free from bubbles and other imperfections. Another method is to soak the mold in lighter fluid and set it ablaze. Until the mold is heated up, the bullets won't be any good. You'll have to cast seven or eight runs of bullets before you get any worth using. Once the mold is hot, you'll be in business.

Because of the time needed for bullets to cool before they can be removed from a mold, many casters find they can use two molds at a time to create bullets at twice the speed. Thus, if you need two types of bullets or two calibers, you can create them all at one sitting.

Lead can be melted in the pot at a high initial setting. Once it is melted, you can turn the rheostat down slightly to keep the lead from oxidizing. This will save you lead and cut down on having to scoop off the oxide on top of the lead. A small amount of flux added to the lead will also help prevent the oxide buildup. When you've finished casting bullets, you can either drain the lead out of the pot and cast ingots in an old pan or what have you, or leave the lead in the pot to harden. If you are working with several types of lead alloy, then you'll probably wish to cast ingots.

Resizing the bullets is unnecessary in most calibers since they will be swaged by the barrel. Accuracy will also usually be better if the lead bullet hasn't been resized. The only times resizing is necessary are when you use a very hard alloy that might cause a high chamber pressure if the bullet didn't fit the bore exactly or when you want to seat a gas check on the bullet's base.

Lead bullets should always be lubricated or they'll foul the barrel after just a few shots. Bullets which are to be used in combat or that may be exposed to temperature extremes require a lubricant with a minimum of oil or wax, which can melt and ruin the powder. Follow the manufacturer's directions for how to best use the lubricant.

Almost any type of lead bullet will work well in a revolver, but may not feed in many autoloaders. To feed through these actions, you'll need a round-tipped bullet cast from a hard lead alloy. A good alloy for automatic pistol bullets is used in lead tire-balancing weights. Pistol brass with a cannelure will keep lead bullets from backing into the cases during recoil and chambering. Bullseye and, to some extent, Unique powders, which are excellent with jacketed bullets, often tend to melt the bases of lead bullets if they are used in full loads. Consequently, you may wish to use gas checks with these powders or use them only in medium to light loads, with a slower burning powder for full charges.

Basically, this is all there is to bullet casting, but it takes a lot of practice to get it right. Bullet casting is more of an art than a science. Though not ideal for combat, the lead bullet can offer a lot of good, cheap practice, and practice is one endeavor that will help you survive an armed confrontation.

SWAGING BULLETS

Swaging bullets is a lot of fun and can create bullets that you cannot find anywhere else. Since no heat is used in swaging, you can create bullets without the dangers of lead fumes, and you can also create jacketed bullets. Unlike the caster, the swager isn't limited by velocity.

A number of books go into great detail on varieties of bullet that can be swaged. I'd suggest that you purchase *Discover Swaging* by David R. Corbin, available from Corbin Manufacturing, and learn what is possible in the way of creating your own bullets.

There are basically two steps in swaging a bullet. The first is cutting or molding lead and placing it into a jacket. This requires a mold or lead wire, cutters and one swaging die. The second operation is shaping the bullet; this is carried out with the second die, or a die set for more complex bullet designs. The bullet's shape is determined by the second die and its weight by the amount of lead added to the jacket in the first operation. Swaging dies generally have a push rod on their tops so that the bullets can be knocked out of the die—they tend to stick because of the enormous pressures involved in forming the projectile.

A lot of pressure is needed to make metal run to fill the dies. For pistol and smaller rifle calibers, a

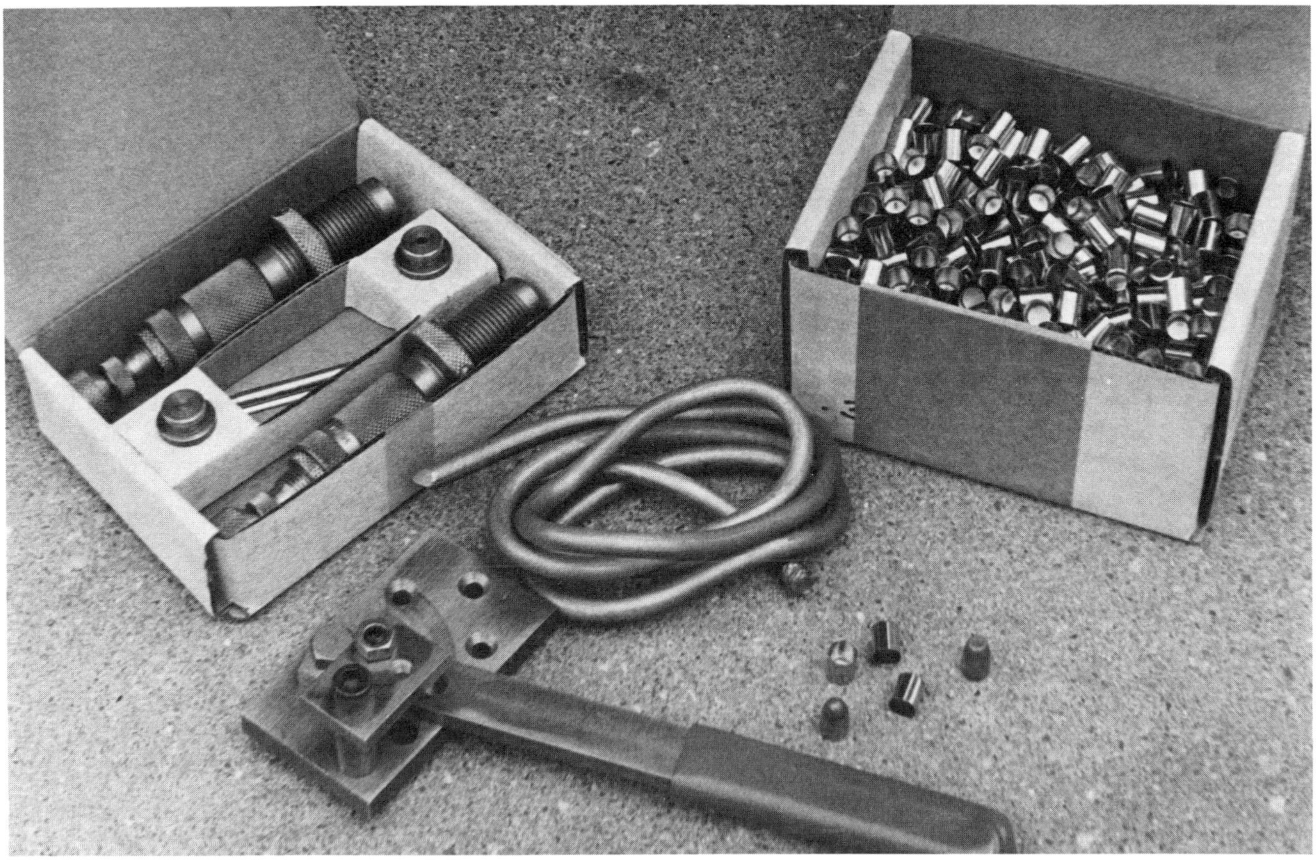

C-H offers a reloader the tools needed to design and make his own combat bullets. Shown above are the C-H swaging die set, core cutter, lead-core wire, and a box of copper jackets.

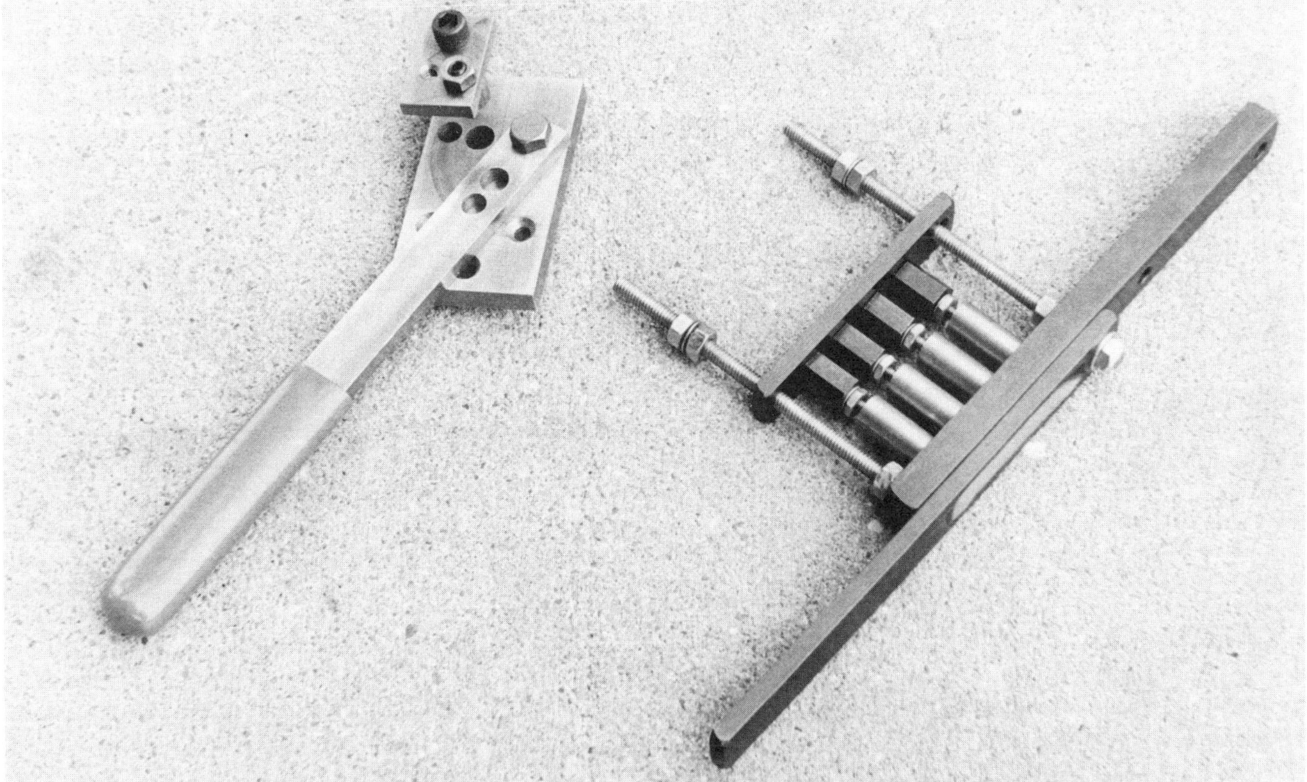

Two solutions to creating lead cores for swaged bullets: the C-H wire core cutter (left) and the Corbin adjustable core mold (right).

hand-loading press can handle this pressure without damage if pure lead is used. Therefore, it's possible to get into swaging with just a few dies, lead wire, lead cutters, and some copper jackets. Try to adjust the die so that the work of shaping is done at the extreme end of the reloading press's swing, where you have maximum leverage. In the next chapter, we'll look at some of the special bullets that can be created by swaging.

YOUR WORK PLACE

You *can* reload at the kitchen table, but it's a better idea to have a tough bench that can take the abuse of having a reloading press bolted into it. If you plan on swaging or melting lead for bullets or bullet cores, then you must find an area with good ventilation well away from food storage areas.

So you will need a heavy bench and a safe place to work. Even if you only clear off the shop table once in awhile and have to bolt and unbolt your press every time, it's a lot better than trying to carry out reloading operations where the family cat or children will help you make mistakes.

A shop has an added benefit in that you'll discover that you need wrenches, pliers, screwdrivers, and other tools while you're reloading. Having the tools nearby is a real ulcer preventer. And again—*don't forget to wear safety goggles.*

5. "Simple" Combat Rounds for Pistols and Submachine Guns

As mentioned before, manufacturers produce extremely reliable ammunition, and most combat experts recommend that you use factory loads. However, there are some types of ammunition that you can only get if you make them yourself. Many shooters who reload do so because it enables them to make better-than-factory loads.

We'll look at modifying commercial ammunition (or reloads) in Chapter 7.

This chapter will cover some powder/bullet combinations that have worked well for me and others. Study these loads to see what can be done and then back off the powder charges a bit when you try them out until you're sure that they won't be maximum loads in your own firearm. If you use military brass, be sure to back off the loads shown by at least one grain; the thicker brass in many of these cases will raise the pressure levels. Let me review the signs of excessive pressure so that you know what to look for: a primer that is flattened or cratered more than usual; primers that rupture or leak; primer pockets that are overenlarged; brass that has "run" or melted into the extractor area; and hard extraction or failure to cycle.

When reloading, try to change only one thing at a time. Finding out that a round shoots better with a different bullet and a different primer doesn't tell which one is better. Try out bullets of similar weight and shape with identical powder and primers. When you find the ideal bullet, try out different powders. Once that's done, try different brands of primer. The end result will be a combination that will give you peak reliability and accuracy in your weapon.

Finally, don't reload or alter ammunition unless you're going to give it your best work. Good reloads will be better than much factory ammunition only if you take the time to develop ideal loads for your weapon and you do your best. Don't lose sight of what you're trying to create: a reliable, lethal round.

ROUND REQUIREMENTS

A defense round in handgun calibers needs to do a lot of things to be effective: its bullet should be capable of penetrating sheet steel, glass, wood, or walls at antagonists who have taken refuge behind such concealment; it should be capable of stopping an opponent almost instantly; and the bullet must expend most of its energy inside an opponent's torso. Any one of these three needs can be easily met. Meeting all three in one round is not so easy.

If a bullet is to fulfill these three tasks, it must be propelled at sufficient speed to give it a high level of kinetic energy, and it must have a configuration that will cause it to expend its energy at a moderate rate. Modern ammunition usually achieves these ends with a lighter than average bullet with a soft or hollow point. The added energy created by its speed allows the bullet to punch through thin barriers while retaining enough energy to expand on impact. There is a thin line here, though. Too much energy will result in over-penetration with the round exiting an antagonist's body or going clear through a house, so that it remains dangerous to bystanders. Bullets which are underpowered or which overexpand cannot be used effectively to penetrate a barrier.

I must stress that having an ideal round is just part of the fight. Semiautomatic and selective-fire weapons are not manufactured to function reliably with anything but FMJ bullets—the MAC-10 family of weapons is especially choosy this way. If at all possible, use a firearm that can reliably chamber rounds with expanding bullets or have your weapon reworked by a skilled gunsmith. This work should include polishing the feed ramp and throating the barrel so that wide-tipped expanding bullets won't get hung up during loading and chambering.

The speed of a bullet and the energy it creates are of no great use if the bullet fails to expand. Some bullets, even though they travel over 800 fps (the minimum speed needed to cause reliable expansion) and are sold as expanding bullets, do not expand on impact. The only way you can be sure you have a reliable load is to test it.

TESTING THE ROUND

Obviously you can't go out and get into a firefight in order to check whether your bullets will expand, so you'll have to experiment a bit. One way to do this is to use ballistic-gelatin blocks, which are very similar to human tissue. They are, however, a chore to work with, and are messy and expensive. An inexpensive substitute is wet newspapers, which give about the same expansion results as the more expensive gelatin (or human flesh).

Believe it or not, you can actually screw up making wet newspapers; if they're not soaked thoroughly for at least twelve hours, they will cause bullets to expand abnormally. You'll need a pile about ten inches thick. To test-fire a round, fire the bullet into the paper while the water is still draining out. The results are only accurate if the paper has a lot of water trapped in it.

Once you've finished shooting, start removing layers of newspaper to see if and how the bullet expanded. The holes and pulping of the newsprint represent the wound channel. Measure the pile from the bottom to see how "deep" into the "wound channel" you're looking.

The bullet should penetrate at least 5-1/2 inches into the newspaper before expanding in order to consistently reach vital organs with a body hit. The bullet should also expand to the point where it is stopped and only fragments penetrate beyond nine inches deep. If the round goes through ten inches of paper, it is overly powerful or didn't expand properly. If it did expand, creating a lot of damage while going through all the paper, remember that in a gunfight, the bullet could exit the bad guy and harm someone behind him. While the newsprint test will not give a very accurate idea of actual wound channels because it doesn't differentiate between the temporary shock-wave wound channel and the permanent one, it will show whether or not a bullet is capable of expanding.

MUZZLE VELOCITY

Once you have determined that a bullet will expand, do some figuring to see how effective it will be in combat. One important factor is muzzle velocity. Ideally you'd have a chronograph to find the exact speed of the bullet, but the chronograph is an expensive tool and a hassle to operate. Fortunately, this information can also be obtained by using the manufacturer's speed data (be sure to allow for velocity data differences if you have a shorter-than-standard-length barrel on your firearm) or the ballistic tables in many reloading manuals. Once you know that the bullet will expand and what its velocity is, you are ready to do some paperwork to figure its combat potential.

The hard work has already been done for us in the various studies mentioned in Chapter 1. While new ammunition has come out since these studies were done, the studies give a realistic idea of what different loads can do. Though a lot of .45 ACP fans are horrified at the terrible showing of their pet FMJ round, the scale does reflect what can be expected in real life given the many variables of bullet placement, body types, etc. These studies also showed why 9mm FMJ did so poorly in combat; it caused very minor disruption in the target and, like the .45, often penetrated without dumping its energy. A quick look at a few of these rounds can give a shooter a good idea how well or poorly his weapon can perform in combat given good or bad rounds of ammunition.

RELATIVE INCAPACITATION RATING SYSTEM

An Evaluation of Police Handgun Ammunition gives a useful arbitrary rating for comparing rounds. Its RI (Relative Incapacitation) rating system covered a range from a high of 54.9 to a low of 0.4. Ratings are based on the amount of disruption caused to ballistic gelatin and no points were given for ease of control, recovery time between shots, etc. Here are a few of the rounds listed and their RI ratings from the report (the first two highest entries for each caliber are given):

RI RATINGS

Caliber	Weight (grains)	Bullet Type	Manufacturer	RI Index
.44 Magnum	200	JHP	Speer	54.9
9mm	96	Safety Slug	Deadeye	54.5
.41 Magnum	210	JSP	Remington	51.9
.357 Magnum	96	Safety Slug	Deadeye	50.0
.44 Magnum	240	SWC	Winchester	50.0
.357 Magnum	125	JHP	Speer	44.4
.38 Special	96	Safety Slug	Deadeye	41.8
9mm	115	JHP	Remington	38.0
.38 Special (+P)	95	JHP	Remington	28.0
.45 Auto	185	JHP	Remington	21.1
.45 Auto	185	WC	Remington	14.7
.41 Magnum	210	SWC	Remington	13.7

A little study of the highest figures shows how important an expanding bullet is. The figures also show that a lighter bullet traveling at a higher speed is more effective for any given caliber than a heavier bullet traveling at a lower speed.

Various nonexpanding .38 Special rounds filled the lowest positions, though the new +P loads hadn't become widely available and were represented by only one load. Of special note is how the FMJ bullets rated compared to their counterparts in the upper end of the RI Index. Here are the low scores for each caliber:

RI RATINGS

Caliber	Weight (grains)	Bullet Type	RI Index
.45 Auto	240	FMJ	6.5–6.7
.44 Magnum	240	SWC	48.9–50.0
.357 Magnum	158	SWC	9.3–10.2
.38 Special	158	Lubaloy	4.1–4.2
9mm	124–125	FMJ and JSP	9.9–13.8

Study of these low figures shows that the .44 Magnum had a high rating even at the low end of the scale. Though one has to wonder why the .45 ACP has been so overrated all these years, the chart makes it obvious why the 9mm has been selected to replace the .45 by the U.S. military, which is limited to the use of FMJ bullets.

Although it doesn't include information on some of the ammunition used in the original 1975 study (including the Glaser Safety Slug), a new 1984 final report reprint entitled *Police Handgun Ammunition: Incapacitation Effects* (available from Paladin Press) shows the various ratings of a number of commercial rounds. It also contains a formula (for which you may need a calculator) for determining the potential of a round *without* having to fire it into ballistic gelatin. The formula is fairly accurate and saves a lot of work in selecting a battle round. (You should still test out the round, however, to be sure the bullet actually expands as it is supposed to.)

Of interest, too, in this final report is the fact that a point of aim centered between an opponent's armpits seems to give a shooter a better chance of scoring a lethal hit than does the older, slightly lower point of aim.

POWER INDEX RATING

A method of calculating bullet performance, using less mathematics, has been developed by Edward A. Matunas and is known as the Power Index Rating (PIR). This system is easy to use and comes very close to actual test results. While the basic formula looks complex, it in fact only involves substituting a few numbers for letters in the formula and then doing a series of multiplications and one division.

Here's Matunas' formula (don't let it scare you):

$$PIR = (V^2 \times E \times B / 12111) \times D$$

The "x's" are multiplication signs, and the "2" means that the number that is "V" in the formula is multiplied by itself (squared). The "/" means that the result of V squared times E times B should be divided by 12111. When all this is done, the result is multiplied by whatever number "D" stands for. It isn't too hard at all.

To find what numbers to plug into the letters, here's what the letters stand for:

PIR = Power Index Rating that we want to find;

V = velocity of the bullet in feet per second;
E = energy transfer (from a table);
B = weight of the bullet in grains;
D = diameter value of the bullet (from table)

To get this formula to work, first you need to find the "E" (energy transfer) value of the bullet in question. Here's how Matunas rates bullets:

- A bullet that expands has an E rating of 0.0100;
- A nonexpanding bullet with a flat nose which equals 60% of the bullet's diameter gets a 0.0085 rating;
- All other nonexpanding bullets get a 0.0075 rating.

Next, you need the "D" (diameter value) rating. Bullets with a diameter (in inches) get the following values:

- 0.2 to 0.249 inches = 0.80
- 0.25 to 0.299 inches = 0.85
- 0.3 to 0.349 inches = 0.90
- 0.35 to 0.399 inches = 1.0
- 0.4 to 0.449 inches = 1.10
- 0.45 to 0.499 inches = 1.15

Let's try one. The .38 Special Federal Nyclad is often used as a defense load. Fired from a 6-inch barrel, it has a velocity (V) of 915 fps. Its weight (B) is 125 grains, and it will expand reliably so it gets an "E" rating of 0.010. The .38 Special uses bullets with a .357-inch diameter, so the caliber is .357 and the "D" is 1.0. If we plug this into the formula, we get as follows:

$$PIR = (915^2 \times 0.010 \times 125 / 12111) \times 1.0$$

Continuing:

$$PIR = (837225 \times 0.01 \times 125 / 12111) \times 1$$

$$PIR = 1046531.2 / 12111 \times 1$$

$$PIR = 86.41$$

These are generally rounded off, so the PIR for the .38 Special Federal Nyclad is 86.

The nice thing about the PIR system is that you can get an idea of how effective a combat load you can develop by reloading and using the ballistic tables available from component manufacturers. Try figuring the PIRs of some of your pet rounds to see how they stack up against others; it can be a real eye-opener. Be sure the rounds really expand if they are supposed to and be sure that you take your firearm's barrel length into account. A 4-inch barrel can suffer a velocity drop of 20 to 80 feet per second compared to its 6-inch counterpart, and a shorter barrel may lose even more. With automatics, barrel lengths tend to be more standardized, with the 9mm Luger chambering having 4-inch barrels and many .45 ACP autos having a 5-inch barrel. With submachine guns with longer barrels, a greater velocity will also result, and semi-auto versions of submachine guns which sport a 16-inch barrel will have an even higher velocity than a pistol or submachine gun with most commercial ammunition.

Once you've cranked all these numbers through the formula, you'll have the PIR rating for your particular load and bullet. This doesn't give you a whole lot until you compare this PIR level to that of other rounds. As a rule of thumb, the following values are generally useful:

PIR of 24 or less: suitable for target shooting or plinking only.

PIR of 25 to 54: good only for small animals; minimal defense use.

PIR of 55 to 94: marginal defense rounds.

PIR of 95 to 150: suitable for military or defense use.

PIR of 151 to 200: suitable for self-defense only with considerable practice to allow for quick recovery from recoil.

PIR of 201 or more: suitable for hunting large animals only; recoil and muzzle blast make these rounds unsuitable for defense use.

A footnote to the PIR: Current research on .22 Long Rifle ammunition suggests that it may be more effective than these figures indicate. Tests with ballistic gelatin show that a .22 bullet starts to tumble three to four inches inside a target and continues to rotate until it travels base first after eight inches. Thus, even if the round fails to hit bone, it will still cause a wound channel larger than the diameter of the bullet. Because of this, shooting victims of the .22 LR are more apt to be put out of action than those shot with a seemingly similar round like the .25 ACP which rarely tumbles. Though the .22 LR is marginal at best, it does better than one might expect with the PIR value system (see Appendix E to see the tumbling properties of the .22 LR in ballistic gelatin).

When playing with the PIR or similar formulas,

Identical ammunition fired from each of these weapons will send out a bullet at different velocities. The ten-inch barrel on this 9mm SGW AR-15 (top) will give higher velocities than the Intratec 99 (center) but lower velocities than the HK-94 (bottom).

it is important to remember that you're coming up with rough figures, not cold facts. Some bullets may expand to a larger diameter than others; some may be more reliable or function better in your pistol. Bullet placement can make a world of difference. A .22 LR placed in an opponent's eye—and brain—would be much more effective than a hit in his finger by the largest caliber available.

GLASER SAFETY SLUG AND SUPER-VEL

There are some very good commercial loads available. One of the best, regardless of caliber, can't be measured with a PIR. This deadly round is the Glaser Safety Slug, a copper shell filled with lead shot. When the bullet enters a target, the shot is released and chews up the tissue through which it passes. These bullets are extremely lethal but have a minimal tendency to ricochet or overpenetrate houses or other structures, which is why they are called safety slugs. Thus, whether you arm yourself with a .25 ACP or a .44 Magnum elephant shooter, the Glaser is usually first choice. Glasers are expensive, but even if you can only afford a few rounds to top off a magazine or put in the first couple of chambers in a wheel gun, the money is well spent. My first recommendation is the Glaser Safety Slug except when you may need the greater penetration offered by lead-core bullets.

In the smaller calibers, Super-Vel is also quite effective. The loads are a little hot, however, so it's generally recommended that they be used only in new steel-framed firearms. The Super-Vel is often equal to, or better than, many of the other loads listed below.

BULLET PREFERENCES

Here are figures for some other rounds that seem to be consistently good (don't forego testing them in your weapon, however. The Glaser Safety Slug or Super-Vel are still usually the first choices in these examples):

ROUND PREFERENCES

Caliber	Second Choice
.25 ACP	Winchester, 45 gr., EXP
.380 Auto	Winchester, 85 gr., STHP
.38 Special	Hydra-Shok, 125 gr., CHHP
.38 Special +P	Remington, 95 gr., JHP
.357 Magnum	Remington, 125 gr., JHP
9mm Luger	Geco-BAT, 86 gr., BHP
.41 Magnum	Winchester, 175 gr., STHP
.44 Magnum	Hydra-Shok, 180 gr., CHHP
.45 ACP	Winchester, 185 gr., STHP

Caliber	Third Choice
.25 ACP	CCI-Lawman, 45 gr., JHP
.380 Auto	CCI-Lawman, 88 gr., JHP
.38 Special	CCI-Lawman, 110 gr., JHP
.38 Special +P	Remington, 110 gr., JHP
.357 Magnum	Winchester, 125 gr., JHP
9mm Luger	Winchester, 115 gr., STHP
.41 Magnum	Federal, 210 gr., JHP
.44 Magnum	Federal, 180 gr., JHP
.45 ACP	CCI-Lawman, 200 gr., JHP

(In this table, "gr." stands for the weight of the round in grains; "EXP" is EXPanding; "STHP" is Silver-Tipped Hollow Point; "JHP" is Jacketed Hollow Point; and "CHHP" is a deeply channeled Copper-Head Hollow Point.)

Winchester's STHP bullets aren't, of course, silver tipped; they have a copper coat with nickel plating, and the jacket has serrations which guarantee that it will peel back as the lead expands (rather than popping off in a ring). These loads are very effective.

The brass hollow point of the Geco-BAT is a hollow point with a plastic insert to give it an FMJ nose shape; a small channel in the base of the BAT blows the cap off the round upon firing and the hollow point is exposed when it strikes the target. The Geco-BAT feeds better than many FMJ bullets and is an effective choice for military arms designed for FMJ ammunition. The bullet is a relatively new design developed to combat terrorism in Europe, where it is known as the "AT," with a high initial velocity enabling it to penetrate as well as or better than armor-piercing bullets. The velocity also creates wounds similar to a hollowpoint, while the lack of density allows the bullet to lose velocity so it doesn't have the long-range danger of standard bullets. The Geco-BAT also has the ability to penetrate some types of body armor and heavier barriers, though it quickly sheds its energy. It is not suitable for long-range engagements, however, because of its quick drop in speed at extreme distances (this is not normally a concern with pistols but might be with submachine guns or carbines).

A round similar to the Geco-BAT which hasn't seen much use in the United States is the French Arcane ("Secret") bullet. Basically, it's a pointed bullet made of copper. Copper is less dense than lead, giving the bullet less weight and more initial velocity. This light weight also causes the bullet's velocity to drop quickly, rendering it less dangerous in urban areas where stray shots can create unwanted harm. Arcane loads are available in .380, 9mm, .357 Magnum, 10mm Bren, .41 Mag., .44 Mag., and .45 ACP. The bullets may be available commercially as well, enabling the reloader to create loads in other similar cartridges.

The 9mm Arcane bullet travels at 1700 fps from a standard-sized pistol barrel. The copper material enables the round to penetrate some armor and ballistic vests, and its speed allows it to create wounds the same way rifle bullets do. The catch is the same as with the BAT: the Arcane bullet loses velocity so quickly that it is suitable only at close to medium combat ranges.

For those who can use expanding bullets, a manufacturer whose rounds perform well in most calibers is Federal. Their JHP ammunition has a very strong following in law-enforcement circles and is often available at discount stores for a bit less than other brands. The JHP ammunition is carefully designed, causes incapacitating wounds, and functions well in most automatic pistols.

Hydra-Shok bullets have an especially deep hollow point with a small column in the center which causes them to expand reliably and quickly. The down side of this design is that the bullets often expand without penetrating very deeply, especially in the small calibers. These rounds are not capable

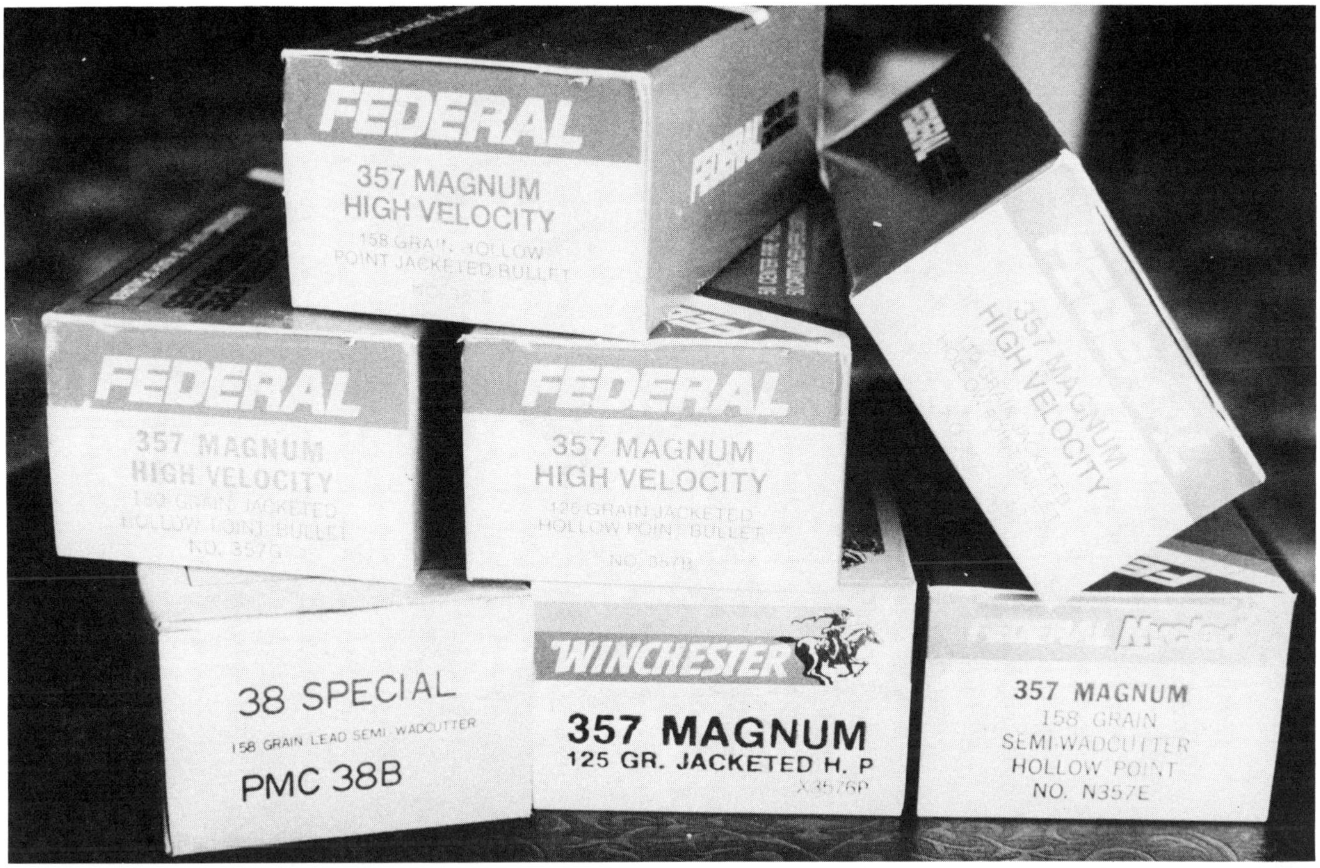

A wide array of reliable commercial ammunition is available for the combat shooter. Shown above are just a few good rounds for the .357 Magnum, including those for combat use and cheap practice.

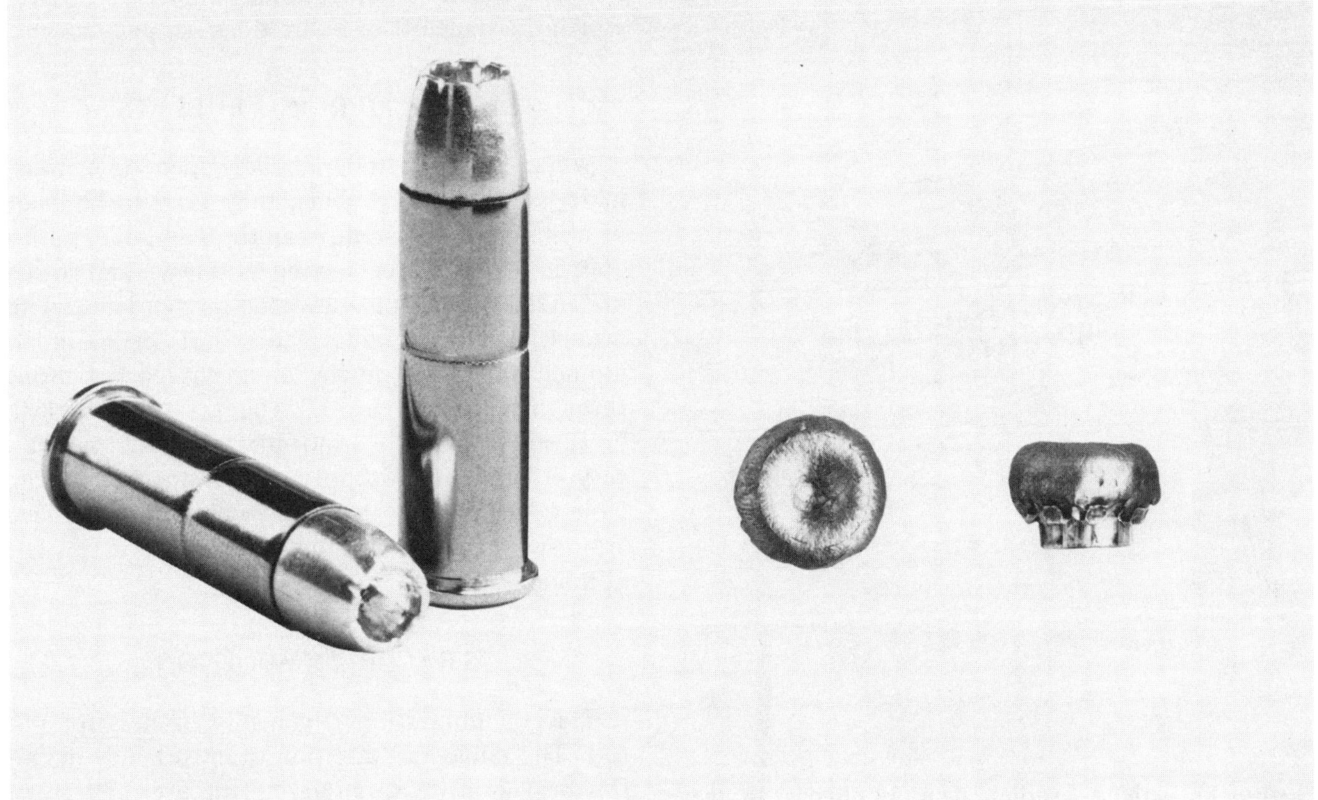

Winchester's "Silvertip" bullet gives both penetration and the expansion needed in a pistol bullet. Photo courtesy of Olin Corporation.

The .41 Magnum Silvertip hollow point from Winchester makes the .41 Magnum a very reliable combat caliber. Photo courtesy of Olin Corporation.

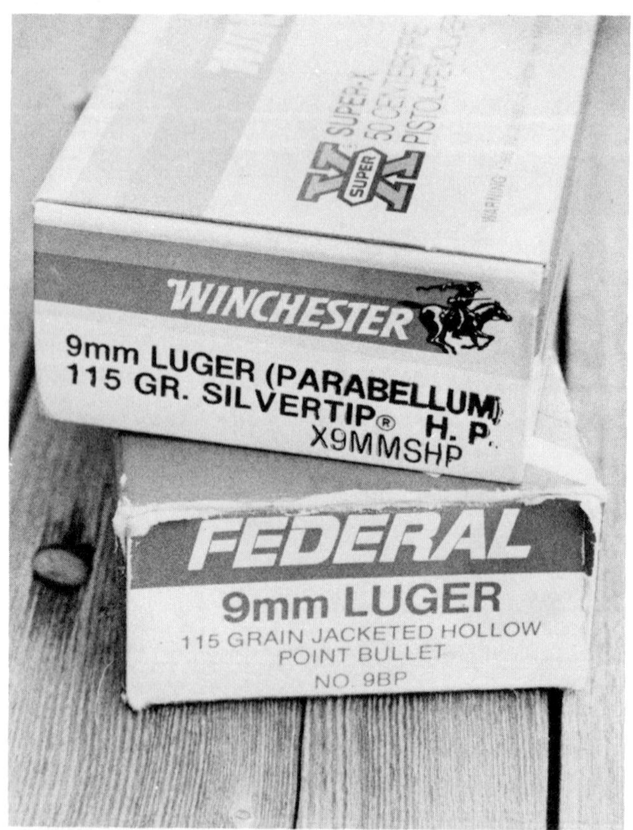

Two of the author's favorite commercial offerings for the 9mm Luger: Winchester's Silvertips (top) and Federal's jacketed hollow point (bottom).

of penetrating heavy cover as well as some other bullets. Better fight-stopping ability might be worth the trade-off of reduced barrier penetration.

ARMOR-PIERCING BULLETS

While armor-piercing bullets penetrate more material than FMJ bullets, they need to be made of a material more rigid than the lead bullet with a copper jacket. Steel, copper, brass, and other materials work, but they aren't very dense. This means that the bullet does not maintain its momentum, and much energy is lost fighting through a tough barrier. The greatest problem, however, is that the armor-piercing bullet is not a very effective man-stopper since it doesn't expand. Unless you know that you will definitely need armor-piercing bullets, they usually are better left at home.

TRACER AMMUNITION

Tracer ammunition is generally useless for pistols. The range and duration of most handgun battles provide little advantage to the user of tracers, especially since they do not come in configurations suitable for combat. (We'll look at tracer ammunition you can make using suitable combat bullets elsewhere.) With submachine guns or rapid-fire carbines, tracers might be of more use. Generally, tracers are used in automatic weapons to aid in getting standard rounds on target, but the high cyclic rate of most modern submachine guns helps out in keeping the point of aim visible.

Many shooters believe they can blow up a car by hitting its gas tank with tracers; in fact, it's nearly impossible to do this, even when fuel is leaking from the tank. Murphy's Law seems to apply in this case: it is only possible to start fires with tracers by accident! Tracers cost nearly a dollar per round. That, coupled with their lack of a good bullet configuration, makes them unsuitable for most combat use in pistols and submachine guns.

EXPLODING BULLETS

Exploding bullets are a very old idea. After their use in the attempted assassination of President Reagan, the .22 version was discontinued, but exploding bullets are still available in most other calibers. These rounds are deep hollow points, in which the cavity is filled with an explosive with a primer of some sort mounted in the tip. When the bullet hits something hard at high speed, the primer ignites the explosive and the bullet fragments. While the explosion itself causes minimal damage, the fragments spread outward at the same speed of the bullet. This creates a multi-projectile wound similar to that of the Glaser Safety Slug. The only problem is that the bullets don't always explode if they don't hit bone or other hard tissue. This unreliability, along with the high cost of the rounds, makes the Glaser Safety Slug a better choice for combat.

PISTOL ROUND RECOMMENDATIONS

From time to time readers will call or write and ask what caliber pistol round I recommend for self-defense. My answer depends on a lot of variables. For one thing, if you're restricted to FMJ ammunition, you've got a tough road ahead. The flatter and wider the bullet tip, the better it is in such a case. Even that isn't the whole story. The old cliché, "One hit is worth a hundred misses," applies. Many shooters can't recover quickly after firing Magnum loads or the .45 ACP. A .38 Special, preferably with a +P load, might actually be a better choice, although with a little training, almost anyone can take the recoil of any round in a standard-sized pistol, with the possible exception of the .44 Magnum.

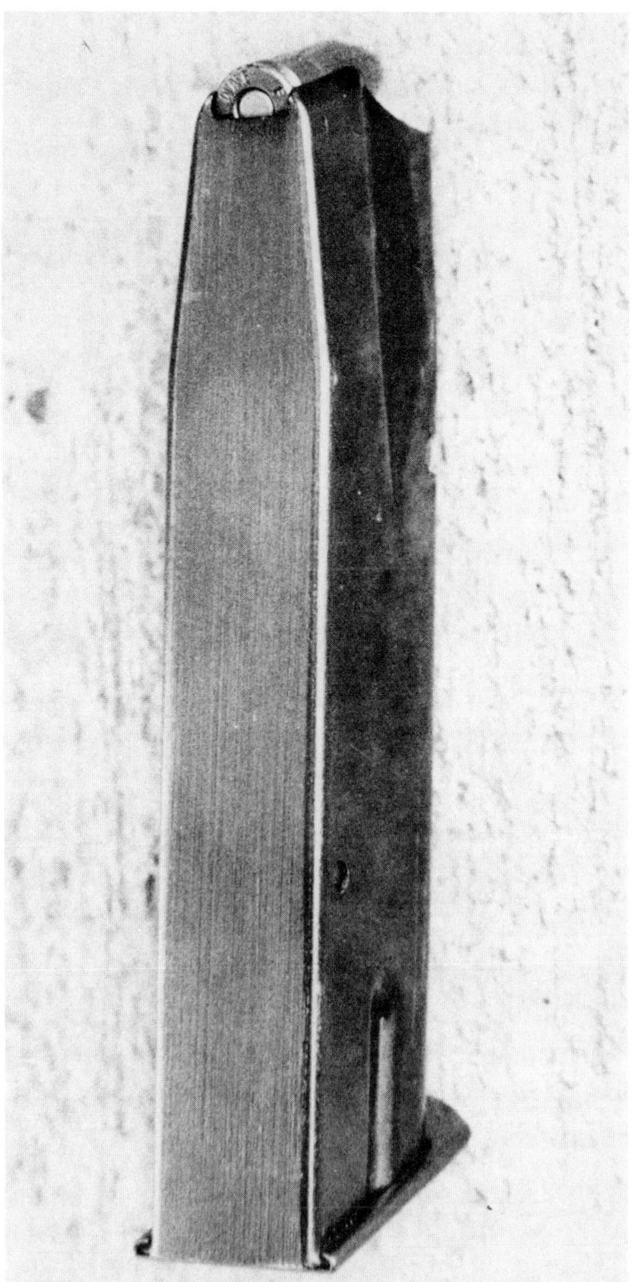

Large-capacity magazines make modern semiauto calibers superior to those designed for revolvers.

The choice also depends on the firearm you carry. In undercover work, having your weapon discovered may be more hazardous than carrying a small caliber pistol. In such a case, the Glaser and Super-Vel loads would be good choice. Too, the high-velocity hollow-point .22 LR often has greater wounding potential than the .32 or even the .380. The flip side is that the .22 cartridge doesn't feed as reliably.

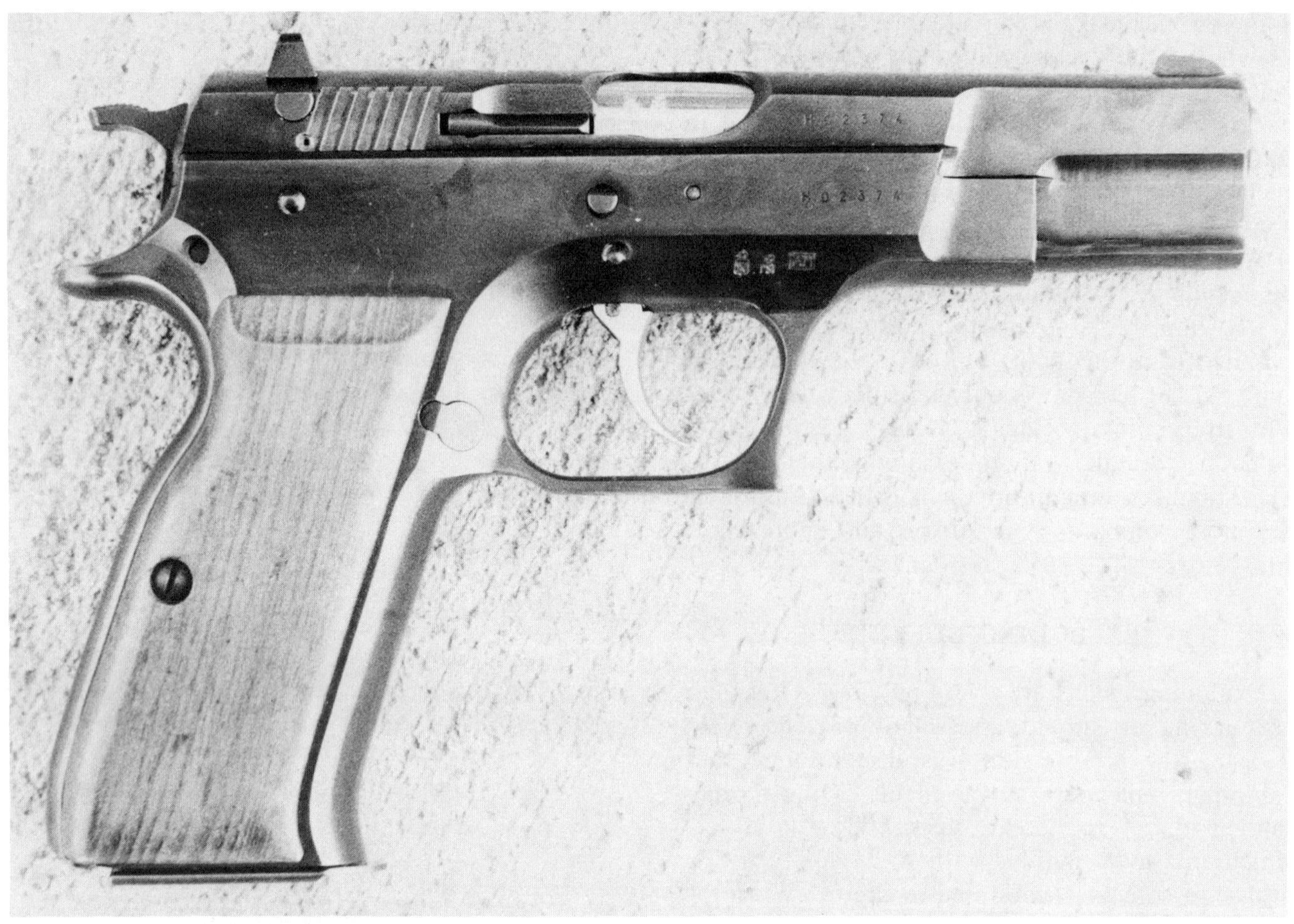

Large-capacity modern pistols, like this TZ-75, plus the superior ballistics of new expanding bullets make the 9mm Luger the author's first choice for a combat pistol round.

If you're not limited in the ammunition you can use, then the 9mm Luger is my all-around first choice for a number of reasons. First, recoil recovery time with an automatic is quicker than for the .357 Magnum, .41 Magnum, .44 Magnum, or .45 ACP; you can fire two or three well-placed shots in the time it takes to place one of the others. Second, magazine capacity is greater in new 9mm pistols. Third, the round is available almost worldwide. And finally, when used in automatic pistols, which are easier to reload and more reliable in dirty environments, the 9mm has greater penetration with modern, expanding ammunition and stopping power equal to any similar round in the .45 ACP.

If we were to run a test with the .38 Special, .357 Magnum, 9mm Luger, and .45 ACP loaded with good expanding bullets, we'd get very predictable results. Each gun would be loaded with the best performing bullets, shown in the following charts, and topped with a standard expanding bullet. Here's what we'd get when we fired into the ballistic gelatin:

The .38 Special (with a +P load) would create a disturbance in about 27 cubic inches of gelatin; the .357 would lead the pack at nearly 50 cubic inches; the 9mm would come in second at 41 cubic inches; and the .45 Auto would be in the last place, slightly behind the .38 +P load, with 26 cubic inches of havoc. The .357 has only a slight edge in such a test. Since the 9mm auto's recoil is so much faster to recover from and automatics are easier to reload, it still gets first place in my choice. If you have to use a revolver, the .357 Magnum is probably the best practical caliber—I don't consider the .44 Magnum practical. But the other large calibers are far from useless and even the .22, .32, and .380 are more deadly than many firearms users realize. And just in case you didn't notice, many of the new +P loads and bullets for the .38 Special give it the wounding ability of the best .45 Auto rounds.

As mentioned above, the 9mm is used worldwide and is standardized, more or less. European ammunition is especially bad about being a little hot when compared to U.S. loads, however. This is not because Europeans have special loads for submachine guns, but rather because they allow a greater gap between chamber and rifling in their 9mm firearms than U.S. manufacturers use. This allows the bullet in a European gun to move forward and relieve the chamber pressure before meeting the resistance of the rifling. The trade-off is that European guns have less inherent accuracy than U.S. weapons. U.S. firearms can safely handle European ammunition, but don't give an alloy-framed pistol a steady diet of hot European fodder.

With submachine guns or pistol-caliber carbines, long-range shots may be fired. To find how effective a round will be at extreme ranges, check a ballistics table to see what speed the bullet retains at various ranges. With this information, it is possible to calculate a bullet's Power Index Rating at extreme ranges. A hollow-point bullet will not deform on impact if its speed drops below 705 to 800 fps, while a soft-point needs a speed of at least 1115 fps to start spreading. Even if these minimums are maintained, expansion will gradually drop off over a bullet's travel so that it expands less and less with greater range.

Regardless of the target distance, submachine guns or carbines that use blow-back operation should not be loaded with the new aluminum-cased "Blazer" ammunition. The lack of support these guns give the aluminum case can cause it to rupture and shoot gas and metal out the ejection port or into the firearm's mechanism.

PISTOL-CALIBER RELOADS

I don't recommend that you use the .25 ACP, .380 ACP, or .32 ACP for self-defense. If you do carry them, the commercial rounds listed above are a better bet than most reloads since there isn't a wide variety of bullet types available in these calibers.

At the end of each load is the energy (in foot-pounds) realized by the bullet as it leaves the muzzle of a standard-length firearm designed for the round. The energy levels are useful for comparing different loads in the same group of cartridges. Don't try to compare the energy level of one type of ammunition to another; it doesn't work out. For example, comparing the highest level of energy of the .45 Auto to the 9mm Luger will show that the .45 has an edge, but the edge is enjoyed only on paper. When the two rounds are tested in ballistic gelatin, the 9mm wins out by a wide margin with modern expanding bullets.

The speeds of the various bullet loads are also given. Generally, these tend to be better indicators of combat performance than energy levels. A fast-moving, lighter bullet creates greater wounds than a slow-moving heavier bullet in the same caliber. In the interest of simplicity and because of the many variables involved in measurement, the bullet speeds have been rounded off.

Finally, a reminder: these loads are safe in the pistols used in testing but are on the hot end of the scale. You should try the combat loads reduced by at least a grain at first and feed your gun a diet of practical loads with only an occasional dose of combat rounds. Firearms are machines that will wear out; the less intense their use, the longer they will serve you.

.38 SPECIAL

Without +P loads, this round is anemic compared to other combat pistol rounds and should be considered marginal. The Glaser Safety Slug has so much going for it that it might be well to use it for combat and reload only for practice. Bullseye and Unique are probably the best powders for accuracy, with HS-6 giving a bit more velocity with most bullets. Flat-tipped, hollow-point bullets are the most lethal, with the lighter-weight versions giving the best combat performance.

Despite its name, this round uses a jacketed bullet that is .357 inches in diameter. Firearms chambered for the .357 Magnum will also chamber and fire the .38 Special (the reverse is not true). The .38 Special is sometimes called the .38 Smith & Wesson Special or the .38 Colt Special, but it is *not* the same as the .38 Smith & Wesson.

Bullets designed for the 9mm Luger can be used in the .38 Special in a pinch. Nine-millimeter hollow points give velocity and accuracy as good as that of a normal .38 Special bullet of the same weight. This is not the most ideal situation, however, because many standard resizing dies don't reduce the case mouth enough to hold the 9mm bullet tightly. Adding cannelures to the bullets and tightly crimping the brass generally take care of this problem.

Loads (with standard .38 Special bullets) are shown in order of combat preference.

.38 SPECIAL JACKETED HP

Bullet (grains)	Powder (grains)	Velocity (fps)	Energy (fpe)
110	Herco (7.6)	1150	323
110	HS-5 (8.4)	1100	296
110	HS-6 (8.4)	1100	296
110	Unique (6.5)	1050	269
110	Bullseye (4.6)	950	220
125	Herco (7.1)	1050	306

With newer, stronger pistols, "+P" loads may be used. While a steady diet of these might cause a .38 pistol to wear out a bit faster than normal, such rounds greatly enhance the combat usefulness of the .38 Special. Provided your pistol can withstand the occasional use of these loads, they can give you an added edge in combat.

.38 SPECIAL +P JACKETED HP

Bullet (grains)	Powder (grains)	Velocity (fps)	Energy (fpe)
110	Herco (7.9–Max. load)	1200	352
110	HS-6 (8.8–Max. load)	1150	323
110	Win 231 (6.5–Max. load)	1150	323
110	Unique (7.0–Max. load)	1100	296
110	Bullseye (5.1)	1050	269

9MM LUGER (PARABELLUM)

The 9mm Luger, also called the 9mm Parabellum, is an old round. It was developed in 1902 for the Luger automatic pistol. It became almost instantly popular in Europe but until recently has had limited popularity in the U.S. Now that the U.S. Army has adopted the round, its place with other combat rounds seems assured.

Early FMJ bullets designed for the 9mm were very poor in combat. Because of this, the cartridge has had a bad reputation as a combat round. Make no mistake: the 9mm *is* poor with FMJ bullets. So are all other pistol rounds. But because of the 9mm's energy, it is a very good combat round with expanding bullets. An added benefit is that the round is small and designed for use in an autoloading action; a number of modern pistols have been designed for the 9mm Luger which give a large number of shots before the need to change magazines arises. A wealth of submachine guns, which also enjoy large-capacity magazines, have been designed for the 9mm as well.

Another plus is that the 9mm Luger is flat-shooting as pistol rounds go. With a 25-meter zero, bullet drop at 50 meters is only 9/10 inch, and at 100 yards, the drop is around 9 inches.

As mentioned before, the 9mm Luger obtains added velocity in submachine guns and carbines. Though military ammunition enjoys only a slight incease in velocity with longer barrel length, many reloads will have far greater velocity with longer barrels if slower-burning powders are used. The user of a carbine or submachine gun would do well to consult several reloading manuals and select one of the slower-burning powders which will give his ammunition a slight edge in velocity.

When reloading the cartridge, care must be taken to fully insert the case into the resizing die. Failure to do so can create chambering problems because the 9mm case is slightly tapered and a partially resized cartridge will not fit into many chambers.

The expander on the reloading dies should be used at a minimum depth so that it flares the cartridge mouth only enough to give the clearance needed to seat the bullet. (Be sure to lightly chamfer the inside edge.) This will help maintain a tight bullet fit so that it won't slip in the brass. Check the bullets for tight fit by holding the brass and pushing the bullet against the reloading bench; if it slips, start over. A loose bullet can slide back during chambering in an autoloader. The 9mm Luger has such high internal pressures even with normal loads that a bullet that slips back will create dangerously excessive pressures. The bullet should withstand at least 35 pounds of push without slipping into the brass.

You may encounter a batch of brass with walls so thin that it is impossible to seat the bullets tightly. If this occurs, it is probably best to discard the brass, although you can make do by rolling a cannelure into the case or by grinding some metal off the expander ball of the die. At the time of this writing, Winchester and PMC brass made prior to 1984 have very thin cases. On submachine guns or assault pistols with blow-back operation, thin-walled brass may actually suffer from blowouts when the brass is unsupported around the head as the bolt cycles backward under recoil. While this occurs only rarely, if you're using such a weapon, it's a good idea to avoid thin brass altogether.

Another way pressures can become too high is if the bullet is seated too deeply. Therefore, keep the overall length of the reload at its maximum; just be sure that it will feed through a magazine.

Though both are chambered for the 9mm Luger, the HK-94 carbine (top) will give bullets a higher velocity because of its sixteen-inch barrel. The pistol shown is the HK-P9S.

This will also help to prevent excessive pressure within the cartridge. Don't take any chances.

For general reloading of pistol rounds, Unique is probably the best powder while Hercules Herco, Winchester 231, Blue Dot, and Bullseye are generally nearly as good. For practice loads using lead-alloy bullets, do not reload for speeds over 1000 fps; leading results above that speed.

Good combat loads in order of preference:

9MM LUGER JACKETED HP

Bullet (grains)	Powder (grains)	Velocity (fps)	Energy (fpe)
90	Unique (7.4)	1400	392
90	Bullseye (5.4)	1350	364
90	Blue Dot (9.5)	1350	364
90	Winchester 231 (6.0)	1350	364
90	HS-6 (7.6)	1300	338
90	Herco (7.7)	1300	338
100	Blue Dot (9.0)	1300	375
115	Blue Dot (8.5)	1200	368
115	Herco (6.8)	1150	368
115	Unique (6.0)	1150	368
115	HS-6 (7.0)	1150	368
115	Win 231 (5.2)	1150	368
115	Bullseye (4.7)	1150	368

Larger bullets are sometimes available for the 9mm Luger. These aren't very good for combat but do work for practice, though the point of impact may be slightly different from the lighter bullets. Check a good reloading manual for possible loads. Here are a few that will probably work well.

9MM LUGER HEAVY-BULLET PRACTICE LOADS

Bullet (grains)	Powder (grains)	Velocity (fps)	Energy (fpe)
124/125	Bullseye (4.1)	1050	305
124/125	Unique (4.4)	1000	277
124/125	Win 231 (4.3)	1000	277
124/125	Herco (4.6)	1000	277
124/125	HS-6 (5.3)	1000	277
124/125	Blue Dot (6.7)	1000	277
130	Bullseye (3.8)	900	234
130	Unique (4.6)	900	234
130	Herco (4.6)	900	234
130	HS-6 (6.9)	900	234
130	Blue Dot (6.3)	900	234
158/160 (lead)	Bullseye (3.0)	900	286
158/160 (lead)	Unique (3.7)	900	286
158/160 (lead)	HS-6 (4.4)	900	286

It is possible to use bullets designed for use in the .38 Special or .357 Magnum in reloading 9mm ammunition. The catch is that these bullets are often too heavy to allow the high velocities that give the 9mm its lethality. They can sometimes be used for cheap practice, however. About the only noticeable problem is that the wider, longer bullets may cause a slight bulging in the 9mm case around the area where the base of the bullet is seated. This doesn't interfere with chambering in most firearms but it does with a few. Avoid the .357-sized bullets for combat rounds and relegate them to practice use only. The lighter bullets can generally be loaded using 9mm data, but use the low end of the tables. Larger bullets can be loaded using the data below provided you are careful to watch for overpressure signs when you first fire the rounds.

Here are some loads designed for use with the heavier .38/.357 bullets.

9MM LUGER HEAVY-BULLET (.357) PRACTICE LOADS

Bullet (grains)	Powder (grains)	Velocity (fps)	Energy (fpe)
140 (jacketed)	Bullseye (4.0)	1010	317
140 (jacketed)	Unique (4.5)	930	269
158 (lead)	Bullseye (3.0)	930	304
158 (lead)	Unique (3.7)	930	304
158 (jacketed)	Bullseye (3.5)	930	304
158 (jacketed)	Unique (4.2)	890	278

.38 COLT SUPER AUTOMATIC

This round was developed from the old .38 Auto and introduced in 1929 by Colt for one of the versions of their automatic pistol. The round has never been popular in the United States or Europe, but a number of the pistols are made by Spanish manufacturers and the round is popular in Canada, South America, and Mexico.

Though slightly more powerful than the 9mm Luger, the .38 Super Auto's lack of popularity and the few good pistols chambered for it make it a rather poor choice for a combat round. The FMJ bullets often available with commercial ammunition make very poor defense loads.

The round is normally loaded with bullets of .357-inch diameter, but it is also possible to use the .355-inch 9mm Luger bullets if no other bullets are available. Care should be taken not to use .38 Super Auto ammunition in a weapon designed for .38 Colt Auto (or .38 ACP); doing so may wreck the weaker weapon.

Unique and Bullseye are probably the powders for this round, though a number of other pistol powders will work with it.

.38 COLT SUPER AUTOMATIC JACKETED HP

Bullet (grains)	Powder (grains)	Velocity (fps)	Energy (fpe)
90	Bullseye (6.7)	1500	450
90	630 (15.3)	1500	450
90	H110 (16.1)	1500	450
110	Unique (6.7)	1300	413
110	Bullseye (5.7)	1300	413
115	Unique (8.0)	1300	413
115	Bullseye (5.9)	1300	413

.357 MAGNUM

The .357 Magnum was introduced by Smith & Wesson in 1935 and is actually a souped up .38 Special with the case lengthened about 1/10 inch to keep the round from being mistakenly loaded into the cylinder of a .38 Special revolver.

The .357 Magnum is powerful enough to be a reliable round provided it has bullets which will expand to take full advantage of its power. It is also hampered by the fact that it must generally be used in revolvers or lever-action carbines. For those limited by choice or regulation to a revolver, the .357 is probably the best choice.

The .357 Magnum round can also be used in carbines, and will gain several hundred feet per second in velocity. Bullet diameter is .357 inch and jacketed bullets are normally used, except in reduced loads, to prevent barrel leading. Unique is probably best suited for reduced loads and target practice, especially with lead bullets. Winchester 296, Blue Dot, Winchester 630, HS-7, and Norma R-123 all work well in combat loads. For reduced target loads, try the .38 Special loads or just use .38 Special ammunition. Occasionally work with full-power loads, however, so that you can get used to controlling the recoil of this round.

Combat loads from barrels shorter than 8 inches will have much lower velocities.

.357 MAGNUM JACKETED HP

Bullet (grains)	Powder (grains)	Velocity (fps)	Energy (fpe)
110	Win 296 (23.6)	1700	706
110	Blue Dot (14.2)	1650	665
110	HS-7 (15.9)	1650	665
110	Win 630 (17.8)	1600	625
110	Norma R-123 (19.4)	1550	587
110	2400 (17.7)	1500	550

BREN 10MM

The Bren 10mm cartridge has a lot going for it. More powerful than either the 9mm or the .45 ACP, it offers a powerful, flat-shooting caliber that will function well in an autoloader. But at the time of this writing, pistols for the round aren't readily available, and many of those that are available appear to have functioning problems. Magazines for the Bren 10 pistol are often in very short supply.

Given the shortage of reloading components and its lack of history, it is possible that so many shooters will hold back and not purchase the pistols, dies, brass, etc., that the Bren 10 will never get off the ground. My advice is to stick with a round that is more readily available.

.41 MAGNUM

Unfortunately, the .41 Magnum has several shortcomings that have been serious enough to limit the round's popularity in combat. Most notable of these are the necessarily large size and weight of a gun that can handle the round, the inherent slowness of reloading, limited capacity of all revolvers, and the fact that the round sometimes opens up too quickly without adequate penetration. Finally, components for the round are sometimes hard to locate. On the plus side is the fact that the .41 Magnum's recoil and overkill make it, from the performance standpoint, a more suitable defense round than the .44 Magnum.

For combat, H-110 and Winchester 630 powders offer the most energy for taking advantage of the round's potential.

.41 MAGNUM JACKETED HP

Bullet (grains)	Powder (grains)	Velocity (fps)	Energy (fpe)
170	AL-8 (18.4)	1450	794
170	H-110 (24.7)	1450	794
170	2400 (21.6)	1450	794
210	H110 (21.0)	1350	850
210	Win 630 (18.7)	1350	850
210	2400 (19.2)	1300	788
210	Win 296 (21.6)	1300	788

.44 SPECIAL

This round was introduced in 1907 by Smith & Wesson. It was based on the black powder .44 Russian. The pistols of the time could not withstand heavy pressures, however, and with the development of the .44 Magnum, the .44 Special has been all but abandoned. The only modern combat pistol made for this round is the Charter Arms Bulldog revolver.

While the .44 can be loaded to give better results than the .38 Special, it is generally a marginal round because its width cuts down on its ability to penetrate cover. That, coupled with the lack of commercial bullets and firearms for the .44 Special, has made this round all but obsolete.

If you have to use the .44 Special, here are some rounds that might work. (A better route is just to use a different cartridge.) Some shooters use much hotter loads and retire their broken pistols from time to time. A better bet is to get a more powerful weapon.

.44 SPECIAL JACKETED HP

Bullet (grains)	Powder (grains)	Velocity (fps)	Energy (fpe)
180	Unique (8.8)	920	338
180	H110 (16.6)	920	338
180	2400 (14.2)	920	338
180	IMR-4227 (18.5)	890	317

.44 MAGNUM

Fielded in 1956 and also known as the .44 Smith & Wesson Magnum or the .44 Remington

Magnum, this is the most powerful commercial round designed for a repeating handgun. This is a dubious honor, however, since the .44 Magnum is hard to control and has enough overkill to make it an unlikely choice as a combat round by anyone other than Dirty Harry. Since it can only be used in a revolver, this caliber is probably one of the last choices among the major defense rounds.

The .44 Special will chamber in weapons designed for the .44 Magnum and makes a good plinking round. Reloading data for the .44 Special can be used in the .44 Magnum, but *not* the other way around. As with other combat rounds, it makes sense to obtain more terminal energy by using a smaller bullet propelled at a higher velocity.

.44 MAGNUM JACKETED HP

Bullet (grains)	Powder (grains)	Velocity (fps)	Energy (fpe)
180	2400 (25.8)	1550	960
180	AL-7 (19.8)	1550	960
180	630 (22.9)	1550	960
180	H110 (28.0)	1550	960
180	296 (30)	1550	960
180	Unique (14.0)	1500	900
180	AL-8 (22.0)	1500	900

Since a steady diet of combat loads is bad for weapons in this caliber, use lighter loads like those below for practice.

.44 MAGNUM PRACTICE ROUNDS

Bullet (grains)	Powder (grains)	Velocity (fps)	Energy (fpe)
240 lead	Bullseye (5.2)	800	341
240 lead	Unique (6.7)	800	341
240 lead	Win 630 (10.9)	800	341
240 lead	Hi-Skor 700X (5.5)	800	341
240 lead	Win 231 (6.5)	800	341
240 lead	SR 4756 (7.6)	800	341

.44 AUTO MAG

This round is a sort of auto counterpart to the .44 Magnum revolver round. The lack of a pistol that stays in production, as well as the lack of factory ammunition, has pretty well made this a collector's item rather than a viable combat alternative.

.45 AUTO

The .45 Auto is the old war-horse of the U.S. military. With expanding bullets, the .45 Auto holds its own pretty well against other rounds and is only slightly less effective than the .357 Magnum and 9mm Luger. The main disadvantage is that the round's large size limits the magazine capacity of pistols chambered for it. Shooters who have the choice would be better off with the 9mm Luger, though one could certainly do worse than carry the .45 Auto for defense. The Colt .45 auto pistol has been around so long that the bugs have been pretty well worked out of it, and a wealth of accessories and modifications are available for it.

Although heavier bullets are popular in this caliber, the lighter 185-grain expanding bullet offers the best combat effectiveness, provided the pistol will chamber it reliably. The recoil of this round has been overrated and is not as bad as many novices would think.

Because of the .45 Auto's popularity in the United States, a wide range of powder and bullet combinations are available for reloading. The best powders from a combat standpoint are Unique, SR 4756, Herco, and HS-6.

.45 AUTO JACKETED HP

Bullet (grains)	Powder (grains)	Velocity (fps)	Energy (fpe)
185	SR 4756 (8.6)	1050	453
185	Unique (8.1)	1000	411
185	Herco (9.2)	1000	411
185	HS-6 (10.3)	1000	411
185	AL-120 (6.3)	950	371
185	Red Dot (6.0)	950	371
185	Win 231 (7.4)	950	371
185	AL-5 (9.6)	950	371
185	Bullseye (6.1)	950	371
200	Red Dot (5.4)	950	401
200	HS-6 (9.2)	950	401
230	HS-6 (9.2)	900	414
230	Unique (6.9)	850	369
230	Herco (7.7)	850	369

Lead bullets can provide cheap practice for the reloader who owns a .45 Auto. The Lyman mold #452460 can be used to cast 200-grain lead bullets.

.45 AUTO LEAD PRACTICE LOADS

Bullet (grains)	Powder (grains)	Velocity (fps)	Energy (fpe)
200	Bullseye (4.4)	750	250
200	Red Dot (4.6)	750	250
200	Win 231 (5.4)	750	250
200	Unique (5.7)	750	250
200	Herco (6.8)	750	250

6. "Simple" Rifle Ammunition

Though not aimed at the combat market, U.S. commercial rifle ammunition has a wide array of rounds suitable for combat. The only reason to reload any of these cartridges would be to obtain ammunition not readily available or for cheap practice. Remington, Winchester, Federal, Norma, and others all offer hunting ammunition which is well suited to combat use. When soft-point spitzers are used, almost any centerfire rifle is capable of lethal hits within normal combat ranges.

With the lower prices for large quantities of commercially reloaded ammunition or inexpensive ammunition like PMC, most shooters may not need to reload, especially those willing to take the time to obtain an FFL. A shooter should think it over before attempting to make his own ammunition unless he needs special bullets such as armor-piercing, tracers, or the like (see Chapter 7).

When picking a cartridge for combat, remember that the high-velocity kill factor begins at 2,000 feet per second. If the bullet drops below this speed, it loses a lot of its lethal potential. This drop can be created by distance or—important to remember when picking out a firearm—barrel length. Most of the figures shown below, as well as in ballistic tables, were obtained using standard-length barrels; if your rifle has a short barrel, take that into account.

With bullets traveling 2000 to 3000 fps, one inch of barrel length will make around a 20 fps velocity change. That means a 10-inch barrel will have a muzzle velocity about 200 fps below its 20-inch counterpart. While that difference may not be important in close combat, at 250 yards or more, it may make or break your rifle's ability for lethal hits.

The West has all but embraced the .223/5.56mm rifle round for combat with the .308 becoming the runner-up. Though the .308 offers more knock-down power with expanding bullets, it is of little value in normal combat because it doesn't start to tumble soon enough. Though it does give extra range, such range is seldom useful in combat. Nevertheless, the .308 is far from obsolete and a number of weapons chambered for it are still found in the hands of very competent warriors. The M1 Carbine and rifles chambered for the .30-06 will occasionally show up in Third World nations, with police sniper teams, or in civilian hands. Though the Carbine's ballistics are terrible and those of the .30-06 little better than the .308, rifles and ammunition are often available for these cartridges when little else is.

The Russian 7.62x39mm round will also often be found in the combat arena. It is generally inferior to all combat rounds except the .30 Carbine. The Russian 7.62x39, the .30 Carbine, the .308, and the .30-06 all use bullets having a diameter of .308-inch. Though different weights of bullets are normally used in each rifle, this allows interchangeability between different rifles with .308 chamberings. While new weapons and rounds are always in the wings, the "Big-Five" (.223/5.56, Russian 7.62x 39mm, .30 Carbine, .308, and .30-06) promise to

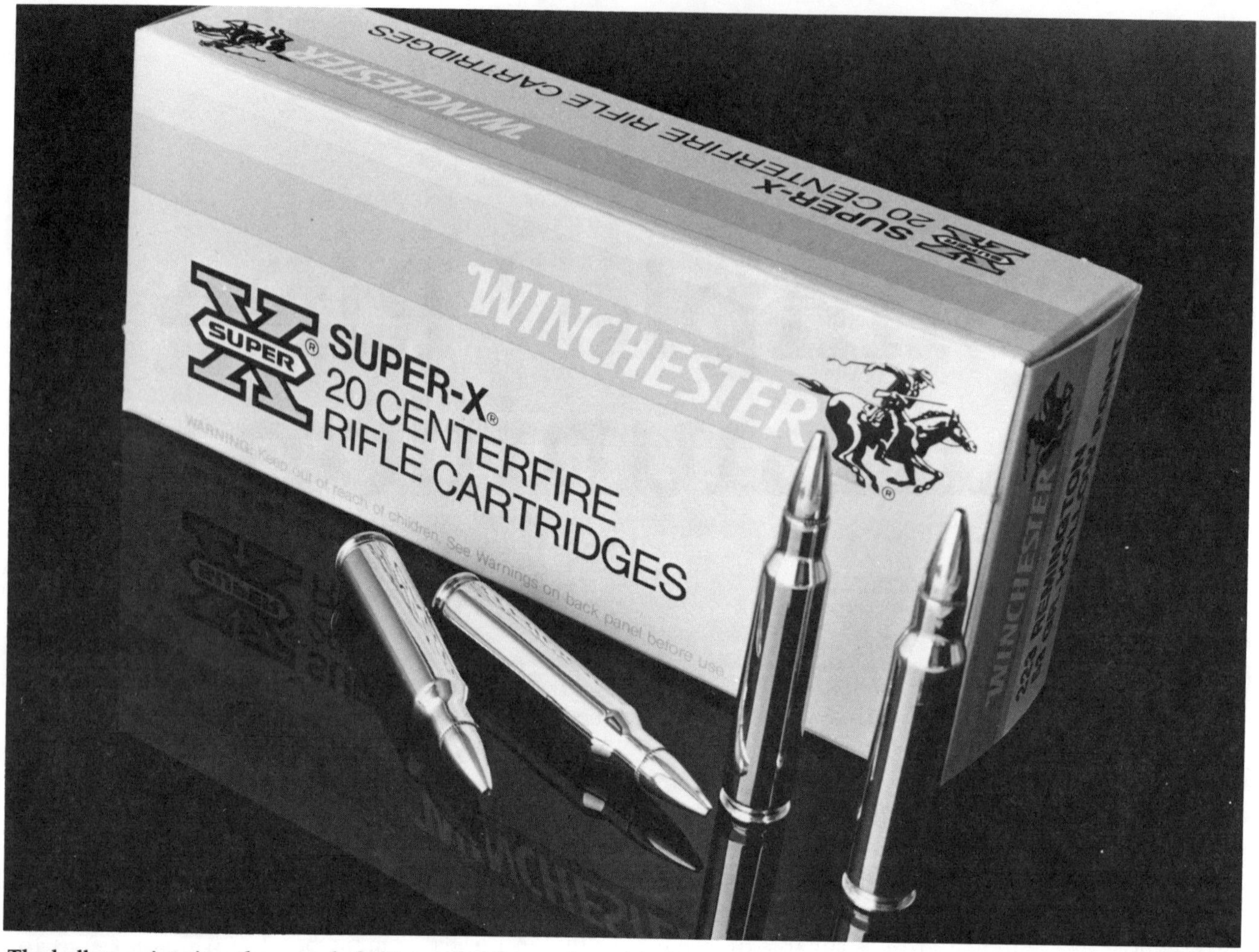

The hollow point gives the extra lethality needed for combat with the .223 Remington. Photo courtesy of Olin Corporation.

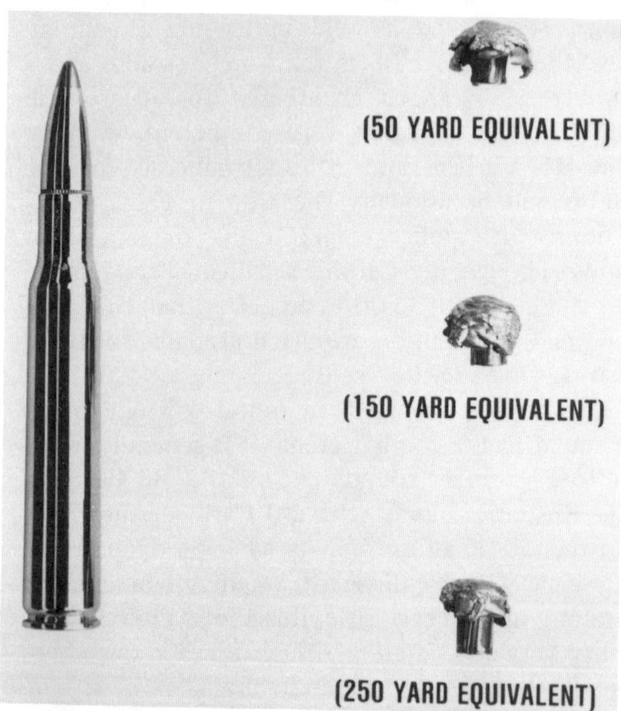

The Winchester commercial round in 30-06 gives a combatant an extra edge within normal combat ranges. Photo courtesy of Olin Corporation.

create havoc well into the twenty-first century and will no doubt defeat many "superior" weapons in future conflicts.

When reloading for rifles, it should be remembered that high speed coupled with a long FMJ bullet that tumbles has been the chosen ammunition for those restricted by the laws of war. Those not so restricted should remember that shorter, expanding bullets, also traveling at high speed, are usually far more lethal and preferred whenever they can be used.

In battle rifles, cycling problems can be caused by too weak or too hot a load. A weak load may fail to chamber a round, cock the hammer, or—rarely except with squib loads—extract the cartridge. At the other extreme, hot loads can cause a round to stick to the chamber wall so that the bolt can't cycle back easily and will occasionally even cause the extractor to break or rip the cartridge rim from the brass. Watch for these problems in addition to checking fired brass. If you're not sure whether it is the rifle or the ammunition that is at fault, try another rifle with the same ammunition or different ammunition in your firearm. If you use expanding bullets, be sure they will chamber reliably in your battle rifle. Check, too, to be sure that exposed lead tips aren't being sheared by the rifle's action so that shavings build up in the mechanism. Good crimps and cannelures on the bullets will help prevent a lot of feed failures and minimize damage to the bullet tips. Chamfering the outside of the mouth of reloaded brass will also help the cartridge slide over the rim of the magazine and up into the chamber.

Armor-piercing rounds of the past weren't too effective. They were generally made of steel or brass and quickly lost their velocity over any great distance. Thus, unless you were engaging a target at close range, the bullet didn't penetrate at all. Many of the new armor-piercing rounds have a steel insert in the bullet point, which allows it to maintain its energy while the point does the work of puncturing armor. The new 5.56mm NATO rounds will probably all be built around this principle.

In rifles which stabilize the bullet enough to prevent tumbling, the all-metal, armor-piercing bullet is not overly effective on soft targets (such as enemy soldiers). On the other hand, it may be the only way to punch through a barrier and reach an enemy. Newer armor-piercing bullets with steel insert tips have both wounding and penetrating capabilities. The 5.56mm armor-piercing bullet still tears apart on impact with flesh.

Tracer ammunition is useful with automatic weapons. Usually one tracer is placed for every four standard cartridges in a belt or magazine, though all-tracer loads are sometimes used. The tracers allow the shooter to compensate visually for any error in the firearm's direction. This works well for belt-fed machine guns, but modern magazine-fed rifles have such a high cyclic rate that they generally run out of ammunition about the time a shooter zeroes in on his target. Consequently, the tracer is of limited use except when fighting in an environment so dark that the sights can't be easily seen, though night sights can get around this problem. Leaders who want to direct fire toward a certain target can do so by marking it with tracers while those under him use standard ammunition. Placing two tracers at the bottom of a rifle's magazine will alert the shooter that he is about to run out of ammunition. This is nearly essential with weapons that don't have a bolt hold-open device to show when the weapon is empty.

Tracers are generally expensive and have such poor ballistics that they aren't an effective combat round. They're a tool to help other bullets get there and do the job. If you don't have access to tracer rounds, see Chapter 7 for ways to create tracer effects with standard bullets.

As with pistol rounds, don't make the mistake of comparing different calibers of rifle cartridges by the energy they generate; it doesn't reflect real life. There's a reason why the .22 has become so popular—it works. It may not look good on paper, but in any realistic test or actual combat, it's very good.

The speeds of the various bullet loads are also given. These tend to be better indicators of combat performance than energy levels. A fast-moving, lighter bullet traveling above 2000 fps creates greater wounds than a slow-moving heavier bullet in the same caliber. And let me repeat the old reminder: these loads are safe in the rifles used in testing but are on the hot end of the scale. You should try the combat loads reduced by at least a grain (more with military brass) at first. And don't forget that your rifle will last longer if you use milder loads for practice.

.30 M1 CARBINE

The .30 M1 Carbine cartridge was developed in 1940 as a replacement for the .45 ACP. In many ways, it is more like a pistol rather than rifle round: it has a straight-walled case and launches a small bullet at low speed. It doesn't come close to being in the same league as the other combat rounds listed below. Nevertheless, it is to be found around the world and there are a zillion carbines squirreled away in the United States.

With expanding bullets, the carbine can be effective at short ranges. Just remember that with FMJ ammunition, this slightly inaccurate weapon is less than marginal, since its normal velocity is below 2000 fps. A lighter .308-caliber bullet would

.30 CARBINE JACKETED ROUND-NOSE (EXPOSED LEAD)

Bullet (grains)	Powder (grains)	Velocity (fps)	Energy (fpe)
77	H110 (17.0)	2400	985
80	H110 (16.0)	2400	1024
80	2400 (16.5)	2200	860
100	2400 (15.5)	2200	1074
110	IMR-4227 (15.0)	2100	1077
110	H110 (14.0)	1900	882
110	2400 (12.8)	1800	791

enhance the carbine's combat effectiveness by boosting its velocity; swagers and hollow-pointers should consider creating 80- to 90-grain bullets for the .30 Carbine if they choose to use this weapon. H110, 2400, and IMR-4227 are the best powders for the .30 Carbine.

It is possible to use lubricated lead bullets with the .30 Carbine. While these bullets are not ideal for combat because of their slightly low velocities and the tendency of the powders to foul the action, such loads can be used in a pinch or for cheap practice, though the shooter must take care to clean the rifle regularly. Generally, it is better to use standard loads and bullets.

Here are several lead bullet loads that may or may not cycle a carbine's action. As with other loads, check the brass for excessive pressure before firing many rounds. The barrel should also be checked regularly for excessive leading. The Lyman #311359 mold may be used to cast 115-grain bullets. Another good mold is the Lee C309-120R for a 120-grain bullet. Harder alloys like wheel weights are better bullet material than softer lead alloys.

.30 CARBINE (LEAD BULLETS)

Bullet (grains)	Powder (grains)	Velocity (fps)	Energy (fpe)
115	H110 (13.0)	1800	828
115	2400 (12.0)	1800	828
120	H110 (14.0)	2000	1066

Extremely light squib loads using lead bullets are also possible. These may or may not cycle the action but can be used for indoor practice with a proper backstop. Don't try jacketed bullets with either of these lead loads. A jacketed bullet will become jammed in the barrel with a reduced load.

.30 CARBINE (LEAD-BULLET) SQUIB LOADS

Bullet (grains)	Powder (grains)	Velocity (fps)	Energy (fpe)
112/115	H-4227 (13.0)	1800	816
112/115	H110 (13.0)	1900	910

7.62x39MM (M43) RUSSIAN

This round was created in 1943 but didn't see much use until after WWII. Though used in the SKS rifle, it is known as the round used in the AK-47 family of rifles. In addition to the USSR, China, and the Warsaw Pact countries, the round has also been adopted by Finland and a number of other nonaligned nations. Though developed from the German 7.92mm Kurz, ballistically the round is quite similar to the U.S. sporting .30-30. In bolt-action rifles chambered for this round, the 150-grain bullet offers extra punch. This heavy a bullet may not work well in some semiauto rifles, however. Lighter bullets (preferably other than FMJ) generally give the best combat performance.

Reloadable 7.62x39mm brass has been hard to find in the United States, but this seems to be changing as more relatively inexpensive AK-47s are imported from China. RCBS currently offers good reloading dies in this chambering. IMR-4198, IMR-4227, 680, and N200 are the most useful powders for reloading with the 7.62x39mm.

7.62X39MM RUSSIAN FMJ, HOLLOW-POINT BULLET OR SPITZER

Bullet (grains)	Powder (grains)	Velocity (fps)	Energy (fpe)
110	IMR-4227 (23.5)	2500	1526
110	680 (24.1)	2500	1526
110	N200 (28.6)	2500	1526
110	IMR-4198 (25.6)	2400	1407
125	BL-C [2] (31.0)	2400	1599
125	IMR-4227 (23.1)	2400	1599
125	680 (23.5)	2400	1599
125	N200 (28.3)	2400	1599
125	IMR-4198 (25.0)	2300	1599
150	IMR-4117 (20.4)	2100	1469
150	680 (20.9)	2100	1469
150	N200 (24.9)	2100	1469
150	IMR-4198 (23.3)	2100	1469

.223 REMINGTON/5.56MM

Probably no round has sparked more heated debate about its use in combat than the .223 Remington. Any problems it may have are greatly reduced for FMJs by either a slow (1-in-12 or 1-in-14) or an extremely fast twist (1-in-7). The slow twist causes the bullet to tumble on impact, while the fast twist causes the bullet to disintegrate on impact, causing a large internal cavern. All doubts about the .223's effectiveness should vanish with the use of hollow-point or exposed-lead spitzer bullets. Because of the round's effectiveness, small size, and low recoil, a modified version with heavier FMJ bullets has been adopted as the second NATO rifle caliber.

Large numbers of sporting rounds use the .224-caliber bullet, so a wealth of bullet sizes and designs are available. SX (varmint) bullets are not suitable for combat and may actually whirl themselves apart in the air with the faster twists sometimes used in .223 rifles. The NATO military bullets may also be too long for some civilian rifles, and should be fired only in military rifles.

The best powders seem to be IMR-4198, H322, and Winchester 748. Many powders and bullets work well with this round; check the literature for what is available. Remember that faster-burning powders will produce less muzzle flash and a slightly higher velocity in the short barrels popular on many combat rifles chambered for the .223 (see the chart in Chapter 4 for powder speeds). Also, the AR-15's gas tube suffers less fouling with faster powders.

In a 1-in-12 twist, the 52- and 53-grain hollow points get peak accuracy with other hollow points following close behind. Most 69-grain or larger bullets will not be stabilized enough for good accuracy if fired in barrel twists less than 1-in-10. A heavy bullet which may be stabilized with a slower twist is Speer's 70-grain Semi-Spitzer Soft Point, which has a less-tapered shape than many other .224 bullets. The trade-off is a decrease in ballistic coefficient, but the power of this soft point gives it greater stopping power than lighter .223 bullets.

Currently, the AR-15/M16A1 rifles have a 1-in-12 twist, the Galil has a 1-in-12, the Ruger Mini-14 has a 1-in-10, and the new AR-15/M16A2 has a 1-in-7. As more shooters see their advantages, newer rifles will undoubtedly be fielded with the faster twists.

.223 REMINGTON/5.56MM HOLLOW-POINT BULLET OR SPITZER

Bullet (grains)	Powder (grains)	Velocity (fps)	Energy (fpe)
40	IMR-4198 (22.6)	3400	1027
40	H335 (27.2)	3400	1027
40	RE-7 (21.8)	3400	1027
40	BL-C [2] (26.5)	3370	1009
45	IMR-4198 (21.5)	3200	1023
45	H335 (27.1)	3200	1023
45	RE-7 (21.9)	3200	1023
45	748 (27.6)	3200	1023
50	IMR-3031 (24.5)	3200	1137
50	748 (27.7)	3200	1137
52/53	RE-7 (20.9)	3100	1120
52/53	IMR-3031 (24.2)	3100	1120
52/53	748 (26.8)	3100	1120
55	IMR-4198 (21.2)	3100	1173
55	IMR-3031 (24.0)	3100	1173
55	IMR-4320 (26.4)	3100	1173
55	RE-7 (21.0)	3000	1099
55	748 (25.9)	3000	1099
55	IMR-4895 (25.1)	3000	1099
55	760 (28.8)	3000	1099
60	748 (25.7)	3000	1199
60	IMR-4320 (25.4)	3000	1199
63	RE-7 (20.5)	2900	1176
63	748 (25.6)	2900	1176
63	H335 (24.4)	2900	1176
63	IMR-4320 (24.9)	2900	1176
63	760 (28.2)	2900	1176
69	748 (26.2)	2800	1201
69	BL-C [2] (22.8)	2600	1036
69	IMR-4895 (24.0)	2600	1036
69	IMR-4320 (24.8)	2600	1036

It is possible to use lubricated lead bullets in .223 rifles. While these are not ideal for combat because of their low velocities, they can be used for cheap practice; care, however, must be taken not to fill the gas tubes of autoloaders with lead shavings. Usually, it's better to use standard loads and bullets or get a .22 LR adapter for practice.

Here are several lead bullet loads that may or may not cycle an auto action. As with other loads, check the brass for excessive pressure before firing many rounds. The use of a harder lead alloy like wheel weights is good for reduced barrel leading. One mold which may be used to cast these bullets is the Lyman #225414, which is listed as a 54-grain mold; with wheel weight alloy the bullet will more likely be 48 grains.

.223 (LEAD BULLETS)

Bullet gr. HP	Powder (grains)	Velocity (fps)	Energy (fpe)
48	IMR-4198 (19.0)	2500	666
48	IMR-4064 (21.0)	2200	516

Extremely light squib loads using lead bullets are also possible. These will not cycle the action but can be used for indoor practice with a proper backstop or even for quiet harassing fire at close ranges. Don't try jacketed bullets with either the above loads or those that follow. A jacketed bullet takes more energy to push through a barrel and will become jammed with a reduced load.

.223 (LEAD BULLET) SQUIB LOADS

Bullet (grains)	Powder (grains)	Velocity (fps)	Energy (fpe)
48	Bullseye (3.5)	1400	209
48	Unique (4.5)	1400	209
48	SR-4756 (5.0)	1400	209

.308 WINCHESTER/7.62MM NATO

The United States started work on a round to replace the .30-06 following World War II. The T-65 was the result, and the U.S. practically brow-beat other NATO countries to choose it as the new standard NATO rifle round. Thus, it became the 7.62mm NATO, while the United States was busy switching over to the .223 Remington—much to the chargrin of her allies. In the meantime, Winchester tacked their name onto the round and released it as a hunting round. Since the ballistics are nearly as good as those of the .30-06 while the case is shorter, the round's popularity seems assured for some time.

Remington offers an "accelerator" cartridge in .308 which uses a 55-grain .224 spitzer bullet in a plastic sabot. Though accuracy is not quite as good as with the standard bullet, it zips out of the barrel at 3770 fps. While it is a little hard to imagine a tactical need for such a load in combat (it's designed for varmint shooting or small-game shooting with the larger .308-caliber rifle), it is good to keep it in mind for special circumstances. Unfortunately, the accelerator bullet and sabot are not available to the reloader.

Because it takes the same size bullet as the .30-06, a wealth of bullets are available for the .308 cartridge. The 168-grain Sierra "MatchKing" is one of the most accurate bullets available for this round and the Speer "Grand Slam" hunting bullets yield a lot of quick knockdown power. The .308 works well with most powders suitable for it and generally hums right along with any of those listed below.

.308 WINCHESTER/7.62MM NATO HOLLOW POINT AND SPIRE POINT

Bullet (grains)	Powder (grains)	Velocity (fps)	Energy (fpe)
110	748 (51.9)	3000	2199
110	BL-C [2] (50.8)	3000	2199
125	748 (48.1)	2900	2335
125	H380 (50.9)	2900	2335
150	748 (48.0)	2800	2612
150	BL-C [2] (46.6)	2800	2612
150	IMR-4064 (46.1)	2800	2612
150	H380 (49.4)	2800	2612
150	760 (53.7)	2800	2612
165	IMR-4320 (44.5)	2600	2477
165	758 (45.3)	2600	2477
165	H335 (44.0)	2600	2477
165	BL-C [2] (44.0)	2600	2477
165	IMR-4064 (43.1)	2600	2477
165	H380 (46.2)	2600	2477
168	N201 (41.0)	2600	2522
168	N202 (44.4)	2600	2522
168	760 (48.5)	2600	2522
168	748 (46.8)	2600	2522
168	H380 (48.2)	2600	2522
180	748 (43.6)	2500	2499
180	BL-C [2] (42.9)	2500	2499
180	IMR-4064 (42.2)	2500	2499
180	H380 (44.7)	2500	2499
180	760 (47.4)	2500	2499
180	IMR-4350 (48.8)	2500	2499
190	760 (47.8)	2500	2637
190	748 (43.4)	2400	2431
190	H380 (45.0)	2400	2431
200	H380 (44.2)	2400	2559
200	748 (43.0)	2400	2559

It is also possible to use lubricated lead bullets in .308 rifles. While they generally fail to cycle an auto action, they can be used for cheap practice. Care must be taken not to fill gas tubes of auto-loaders with lead shavings, and gas checks should be used on the 150- and 180-grain bullets.

As with other loads, check the brass early on for excessive pressure and for excessive leading. Again, wheel weight alloy is a good bullet material. Molds which may be used to cast bullets for the .308 are the Lyman #311455 (152 grains), and #311332 (180 grains).

.308 LEAD BULLET LOADS

Bullet (grains)	Powder (grains)	Velocity (fps)	Energy (fpe)
150	H4831 (43.0)	1900	1203
150	H4895 (30.0)	1800	1079
180	H4895 (29.0)	2000	1599
180	H4198 (23.0)	1800	1295

.30-06

Introduced as a military round in 1906, this cartridge was a modification of the .30-03 made by changing the bullet to 150 grains and shortening the case somewhat. Since its introduction, it has been used as a military round by a large number of countries at one time or another and has had widespread use as a hunting cartridge. Though the round has too much powder for a modern combat round, it is still used in older combat rifles like the M1 Garand and will continue to find use, with expanding bullets, as a sniper load.

Everything the .308 can do, the .30-06 can do just a little better. For some, this makes up for the awkward size of the cartridge. A wide range of bullets and loadings are available for the cartridge including the Remington "Accelerator." The .224 sabot bullet flashes out of the barrel at 4080 fps.

A wide range of bullets is available for the .30-06. The lighter 110-grain bullets should be reserved for long-range use since they tend to overexpand at close ranges, making a massive but sometimes superficial wound. The 150- and 180-grain bullets are generally the most satisfactory, with the 180-grain coming closest to the original 176-grain military loading. When surplus military brass is used, it is a good idea to back off the following loads by at least 1.5 grains to prevent excessive pressures. It's a good idea not to use surplus ammunition for combat; it is old enough to have lost reliability.

.30-06 HOLLOW POINT AND SPIRE POINT

Bullet (grains)	Powder (grains)	Velocity (fps)	Energy (fpe)
110	748 (60.2)	3400	2824
110	N202 (59.1)	3400	2824
110	BL-C [2] (51.9)	3200	2502
125	IMR-4064 (53.2)	3100	2668
125	H380 (54.3)	3100	2668
125	BL-C [2] (52.2)	3100	2668
150	760 (60.6)	3000	2998
150	748 (54.3)	3000	2998
150	H205 (59.8)	3000	2998
150	IMR-4350 (57.5)	2900	2802
150	IMR-4064 (51.2)	2900	2802
150	H380 (51.5)	2900	2802
165	H205 (58.1)	2900	3082
165	760 (58.5)	2900	3082
165	IMR-3031 (46.1)	2700	2672
165	IMR-4064 (47.9)	2700	2672
165	H380 (49.2)	2700	2672
168	IMR-4350 (55.8)	2800	2925
168	H205 (56.6)	2800	2925
168	760 (56.4)	2800	2925
168	748 (51.8)	2800	2925
168	Norma MRP (61.5)	2800	2925
180	785 (61.1)	2700	2914
180	Norma MRP (60.3)	2700	2914
180	IMR-4350 (54.5)	2700	2914
180	IMR-4831 (57.3)	2700	2914
190	760 (55.3)	2700	3076
190	785 (60.0)	2700	3076
190	IMR-4350 (54.8)	2700	3076
200	IMR-4350 (52.5)	2500	2776
200	785 (57.0)	2500	2776
220	IMR-4350 (50.9)	2400	2814
220	H450 (52.0)	2400	2814

As with the other U.S. military rounds, it is possible to use lubricated lead bullets in .30-06 rifles. See the section on the .308 for limitations and precautions. Gas checks should be used on 150- and 180-grain bullets. Several molds which may be used to cast bullets for the .30-06 are the Lyman #311359 (115 grains), #311455 (152 grains), and #311332 (180 grains).

.30-06 LEAD BULLET LOADS

Bullet (grains)	Powder (grains)	Velocity (fps)	Energy (fpe)
115	H4895 (26.0)	1500	575
115	H4198 (23.0)	1800	828
150	H4831 (45.0)	1900	1203
150	H4895 (30.0)	1800	1079
180	H4895 (31.0)	1900	1443
180	H4198 (23.0)	1800	1295

SNIPER ROUNDS

Regardless of the cartridge used for sniper work, consistent accuracy is of prime importance. While sniping usually occurs at normal combat ranges, it may occasionally be necessary to attempt shots at 500 yards or more. Under such conditions, a good handload is often more accurate than factory ammunition, if it's done correctly.

There are several factors involved here. All the components must be virtually identical to place shots consistently. If powder loads, brass walls, bullets, and so on vary from round to round, the ammunition will lack the inherent accuracy sniper rounds need, which can be achieved by uniformity. To this end, the reloader must weigh and inspect each component, and make all motions used in operating the reloading equipment identical to ensure that bullet seating depth, resizing, powder load, trimming, etc., are consistent from one cartridge to the next. For the most reliable loads, a primer pocket reamer that makes all the primer holes of equal size helps eliminate the last fractions of an inch of error.

The really serious sniper may even check the hardness of each brass cartridge with a Rockwell Hardness Tester, normally used in machine shops, to be sure all the brass will stretch and release the bullet in a consistent manner. A hardness of 82 to 87 Rb is said to be the best range for rifle brass.

Finding the strength of the brass yields another plus: by using harder and stronger brass, slightly larger maximum loads are possible. While maximum loads may not be as accurate as others, they give a little extra room for a sniper to experiment when trying for a super-accurate load, which brings up the final criterion for super-accurate loads: experimenting. An accurate load can only be developed by testing various combinations of components and powder charges. A perfectly good load can be extremely inaccurate if bullet seating depth is changed only slightly or if the powder charge is raised or lowered by a grain or less. No book can tell you how your rifle will react with a given load. Try full-length resizing and neck-sizing only; try seating the bullet at different depths; try trimming the cases to different lengths within the maximum/minimum range. The hardest part of developing a sniper round is experimenting. It takes time and is the only way to achieve the accuracy needed for sniper work.

Regardless of the weapon you use or the conditions you fight under, two things will greatly improve your chances of survival: lots of practice and high-quality, reliable ammunition designed to create maximum wounding effects with an expanding bullet.

7. "Advanced" Pistol and Rifle Ammunition

You'd be better off to skip this chapter if you don't need anything other than standard commercial ammunition or simple reloads. It's easy to get caught up in buying or creating special loads that may prove useless in combat. So if you devise a special load for self-defense, be sure it's 100 percent dependable and will do what you want it to. One mistake is to depend on trick bullets to make up for a round's inherent lack of power. The .25 ACP, for example, will never be very effective even with a liquid-filled high-velocity explosive bullet. Likewise, a pistol round will never equal a small rifle cartridge like the .223. Get a short rifle if that is what you really need.

Care must also be taken in developing rounds or trying to duplicate special commercial rounds. Think before you fire a special round in your firearm. If you've created a bullet that leaves its jacket behind in the barrel or something similar, your experiments will get very expensive in a hurry.

If you know the turf on which you'll be fighting or that your enemy may be wearing a ballistic vest, then special bullets can give you a distinct edge over standard ammunition. Even here, though, there is a down side, and that is that specialized ammunition is only good for limited conditions. A specialized load for close combat may be useless for a long-range shot; bullets that cut through ballistic vests may be dangerous to occupants of a nearby apartment. While some argue that you can quickly substitute a magazine of standard ammunition in your weapon, in combat you may have neither the time nor the presence of mind to do so. Likewise the idea of using staggered standard and special rounds often fails to work well—you may forget the sequence. Go with one or the other but don't plan on switching back and forth.

Except for Super-Vel, Geco-BAT, the Glaser Safety Slug, and special bullets from National, multiple-projectile rounds, liquid-filled bullets, etc., come and go. A few never even appear in the maketplace, which stands to reason; how many people do you know who will spend extra money on special loads? A manufacturer has to sell a lot of ammunition to pay the bills, and specialized ammunition—unless it is adopted by the police or military—doesn't sell at the high price it must command to be profitable.

Another reason manufacturers don't market some types of ammunition is the bad publicity that could result. Can you imagine what the anti-gun press would write about a manufacturer who marketed liquid-filled bullets that created massive "cruel" wounds rather than punching neat little "humane" holes? It doesn't take much imagination to see why the already hounded ammunition manufacturers don't care to get out on such bad publicity limbs.

Pressure may eventually force manufacturers to discontinue the few specialized rounds they now offer or limit their sale to the military and police. It's not hard to imagine a time when you might not be able to purchase standard combat rounds, let alone specialized ammunition. In such a case, being able to make rounds suitable for your needs may be the only way you'll ever be able to use it.

Provided you use a little common sense and caution, you can create many useful types of ammunition from standard factory loads without sacrificing reliability and with a bare minimum of reloading equipment. A bullet puller and an inexpensive Lee or Mequon hand-loader kit will allow you to pull the bullets, modify or even replace the bullet, and reassemble the ammunition, but you will be limited to just those operations. If you reload, you'll have a wider range of options open to you. Maximum flexibility will come from purchasing some swaging equipment and possibly a small lathe.

It is easier to create special loads for larger-caliber ammunition because the larger bullets are easier to work with. Because of this—as well as the added lethality of most rifle rounds—pistol bullets are most easily modified for custom loads while rounds like the .223 Remington are harder to work with.

Whatever type of bullet you modify, do so cautiously, as such rounds are extremely dangerous. Use reduced loads until you know the altered bullet will go down the barrel with the same amount of pressure. Don't fire more than one round at a time until you've inspected the barrel to be sure a jacket or other residue hasn't been left behind. When in doubt, follow the most conservative course. You'll save yourself a lot of trouble and money, and maybe even some doctor bills.

TRACERS

Tracers may be illegal in some areas, and a lot of people assume they are illegal in any case, so you may run into hassles from the law even if you're using tracers legally. Check local laws before you start testing them out.

When firing tracers, remember that they can have an incendiary effect, which makes their use in dry areas a definite mistake unless you're trying to start a forest fire.

Currently a few tracers are available on the military surplus market. More reliable tracer ammunition is manufactured by National. Their ammunition leaves a red trace and is available in .38 Spl., .357 Magnum, 9mm, .45 Auto, and .380 ACP; dealer's price on a box of ten cartridges is around $10. Tracers in rifle calibers aren't as easy to obtain on the commercial market, though National would be a good place to check. Because of their price and the difficulty in obtaining them in rifle calibers, you may wish to make your own tracers.

A tracer's illumination comes from chemicals in the base of the hollowed-out bullet, ignited by the powder of the cartridge when it's fired. The burning length of the chemicals can be quite short though generally enough is used to create a trace out to the projectile's full combat range. The more tracer chemical there is in the bullet, the less accurate and lethal it becomes. So if you are using tracers at the normal combat pistol range, you will greatly enhance the round's effectiveness by using only enough tracer chemical in it to cover that range.

It would be possible to purchase tracer ammunition in .308 Winchester and use a collet bullet puller to remove the bullets (don't use a kinetic puller—you could ignite the tracer). This would give you a .30 caliber bullet that could be loaded into .30 Carbine, .30-06, or 7.62x39mm Russian cases as well as a different .308 cartridge with a new powder load. Perhaps you've obtained some old ammunition that doesn't propel the tracer bullets at high speeds. Again, pulling the bullets will allow you to create new tracer ammunition. Since the tracer chemical's brightness will dull only slightly with time (unlike the propellant, which will really fall off), the tracer bullets themselves remain good for a very long time.

When you reload the tracer bullet, first weigh it, then use reloading data to create a round as similar as possible to the ammunition you normally use. This will help the point of impact remain as constant as possible, remembering that the tracer's point of impact will be far off at extreme ranges because of its different ballistic shape and weight.

Another possibility is to alter the bullets so that they have more combat effectiveness. Since tracers are usually FMJs, you'd do well to create soft points, in the manner outlined below.

If you can't buy tracer rounds, then your job is a bit more complex. Your first step is to get the tracer chemicals. The easiest way is to use the filler from a standard highway safety flare. After *carefully* removing the striker cap, you will have a wealth of material to use for tracers. Use a small knife or other tool to carve the powder out of the opening and catch it on a clean piece of paper. If the material is in large chunks, carefully break it apart. Other chemicals which may be used for tracers include barium peroxide, aluminum flakes, magnesium powder, strontium nitrate, strontium peroxide, or combinations of these.

Your next step is to prepare the bullet. A jacketed bullet of the same type you normally use is ideal. It should weigh a bit more than the standard bullets since you'll be digging lead out of it to make room for the tracer chemical; usually the bullet need only be several grains, however.

The closer the hole for the tracer is to the bullet's center, the more accurate the tracer will be. If possible, place the bullet in a metal lathe and turn a small hole in its base. A drill press or even a power drill anchored to a tabletop can also be used. Spin the bullet, use a small tool to dig into the base, and try to make a clean cut. Not a lot of lead has to be cut away, but for best results, the hole should be shallow and wide rather than narrow and deep. The longer the tracer needs to burn, the deeper the hole should be. (There are limits to the length of burn, however, since burnt chemicals in the base of the bullet tend to obscure the glowing material.) Start out by removing just several grains of lead and replacing them with tracer; test-fire to see if you've used enough chemical.

The tracer chemical can be secured in the bullet base in several ways. The best is to dissolve the chemical in a solvent, then place the liquid in the base of the bullet and allow it to dry. An easier method is to use epoxy to set the chemical into place. The easiest to use is "Five Minute" epoxy, which allows the bullets to be moved about a few minutes after the tracer has been placed in the cavity.

After mixing the two epoxy components, add the glue to the tracer chemical until it is just sticky enough to adhere to the cavity in the bullet. The less glue you use, the better the tracer will burn. Shellac can also be used with good results. Pack the glue and tracer mixture into the bullet cavity and set it aside to dry. Once dry, the bullet can be seated in its case, but it should not be fired for twenty-four hours or more so that the glue has time to cure.

Red lacquer is normally placed on the tips of tracer bullets to distinguish them from standard rounds. Fingernail polish works well and comes complete with an applicator brush to simplify your job.

Though tracers will not be overly damaging to a firearm's barrel, it is wise to clean the barrel carefully before extended storage after using tracers, and to avoid a steady diet of the rounds. Normal practice is to use one tracer for every three or four standard rounds. In addition to cutting down on possible barrel wear, the standard rounds will help to remove some of the tracer fouling.

When using a tracer, it should be remembered that the round can be seen by anyone who is off to your side or behind you. Tracers can be very helpful, especially in low-light conditions, but they can also give away your position.

ARMOR-PIERCING BULLETS

The simplest armor-piercing bullet is just a steel, brass, or copper rod turned on a lathe to the correct bullet dimensions. Armor-piercing rounds of this type are still available both as surplus and as new commercial ammunition. One interesting trick is to place a nylon jacket around the metal penetrator. While this doesn't improve its penetration, it does cut down on friction and barrel wear. Probably the best known of these types of rounds is the KTW, which is currently restricted to use by law-enforcement and the military.

The solid-metal penetrator bullet is easy to make, but before you create umpteen of these, be warned that, especially with the larger pistol calibers, they don't penetrate much better than FMJ bullets on steel plate or vehicle armor. They are effective on wood, sheet metal, or other barriers that often break up or deflect regular bullets, however, and in .308 or .30-06 can also do quite a number on a car engine.

Military instructors often like to wow the troops with tales of how 7.62mm rifles will shoot through trees and other barriers. A modern FMJ round will go the distance through wood but is apt to take a turn when it hits the rings of a tree. This is where the solid brass bullet shines. A .308 brass bullet made following the directions below can usually penetrate a 14-inch-diameter tree and still have enough power to penetrate a 1/16-inch mild steel plate. A bullet like that really cuts down on the places a foe can hide.

Standard pistol bullets don't do well in penetrating cars. The .44 Magnum, .357 Magnum, and 9mm will often go through car doors, but can be inconsistent, since there's a lot of excess metal and support rods inside the door. The .45 fired from a pistol will seldom penetrate an automobile door. Interestingly, the Geco-BAT and the French Arcane handgun bullets both display some armor-defeating abilities at close ranges.

In rifle calibers, the .223 is notoriously bad at

penetrating a car door. The problem is that the bullet tends to break apart, showering lead and jacket fragments on those in the car. This tendency makes it lethal on the battlefield but poor at police barricades. While I wouldn't volunteer to sit in a car while someone took potshots at it with a .223 rifle, the round just isn't all that reliable against an opponent in a car. The .30 Carbine is also marginal in this application. The 7.62x39mm Russian, .308, and .30-06 are better, with the two U.S. rounds being especially effective.

Pistol shots to an engine will rarely inflict any quick damage. The .223 may damage an engine if a barrage of rounds is delivered through the front grille. The .30 Carbine is also an iffy proposition here unless a high volume of fire is used, and, strangely enough, so are standard .308 and .30-06 rounds, since they are harder to control when sending the number of shots needed to defeat a car engine. An engine block is pretty tough; only lucky hits make a difference and those are more apt to happen if a large number of shots are fired. The likelihood of one accurate shot stopping a car is very slim.

Things change with armor-piercing bullets. Though the .45 ACP still often doesn't make it through the side of a car when fired from a pistol, the .357 and 9mm are pretty good, as is the .38 Special loaded to +P loads. The .223 AP is an improvement, but still a little shy about going through a car door. The 7.62x39mm, .308, and .30-06 zip right through and are far more likely to damage an engine.

So bullet penetration improves with armor-piercing rounds, but not dramatically. Regardless of caliber, turned-metal bullets tend to lose their speed over any great distance and also have a ballistic arc different from standard rounds. You will have to rezero your sights or adjust your point of aim slightly for full accuracy with these rounds, and try to fire from shorter ranges.

The turned-metal bullet is the easiest to make. A good dial caliper is essential to achieve the correct bullet diameter, and a metal lathe makes the job consistently easier, but a drill press or even a power drill in a stationary mount will also work. The best material to use is probably brass or copper. Steel bullets are hard and will quickly ruin a barrel, or the bullet can lodge in the bore. Brass rod is available in many hardware stores or welding and machining shops. Purchase rod of only slightly greater diameter than the bullet you'll be making; it saves a lot of file work.

Don't try to cast bronze bullets in lead-bullet molds. Brass melts at a much higher temperature, and steel molds aren't made for the temperature needed. Aluminum molds will undoubtedly be damaged.

Cut the rod to several inches in length. Too long a piece will wobble as it spins. Chuck the rod in place, spin it at a moderate to high speed, and use a medium-toothed file to shape and a fine-toothed file to finish it. Check your work with calipers from time to time. While you can resize a cartridge and check the fit of your rod in the case mouth, calipers are a lot safer.

When the rod is the proper diameter (.308 inch for .30 caliber, .224 for .223 Remington, .355 for 9mm, etc.), your next step is to cut small gas relief grooves into the side of the bullet. These cuts keep friction down and reduce barrel wear. Depth of the relief grooves is not critical, but here's a table of suggested diameters for both the bullet and the relief grooves:

BRASS ARMOR-PIERCING BULLET
(Dimensions in inches)

Bullet Diameter	Relief Groove Diameter	Bullet Length
.308 (.30-Caliber rifles)	.250	1.25
.224 (.223 Remington)	.180	0.90
.355 (9mm Luger/.38 Super)	.288	Standard
.357 (.38 Special/.357 Mag.)	.289	Standard
.410 (.41 Magnum)	.332	Standard
.429 (.44 Magnum)	.348	Standard

While the rod is still in the chuck, use the file to cut a point. The shape of the point doesn't seem to be too important to penetration as long as it is tapered and allows the bullet to feed through the weapon's action. With rifle bullets, a streamlined shape with a blunt or even flat point seems to give better penetration when the bullet hits a barrier at an angle. With pistol bullets, a cone-shaped point with a flat tip gives superior results.

When the overall shape is correct, polish the rod with a fine file, then complete the burnishing with steel wool. The rod can be cut to length with a file or a hacksaw blade. Make the cut while the rod rotates in the chuck, but slow the speed down and be sure to wear safety goggles. At this point, the real purist will place the bullet back into the

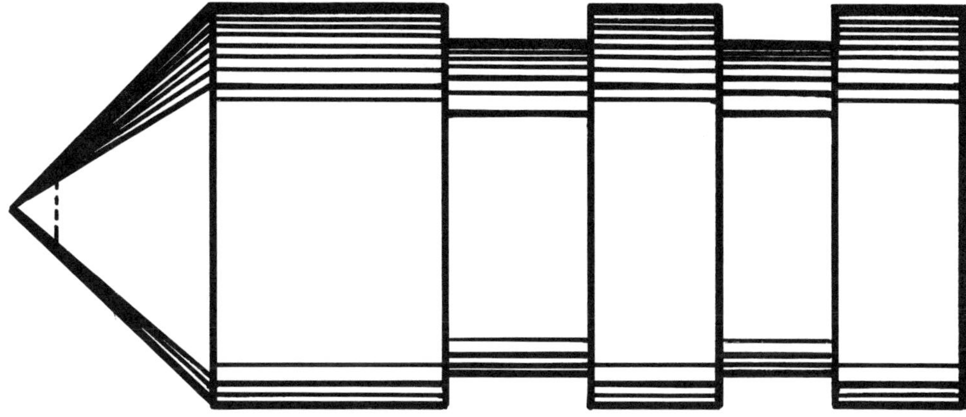

Machined-brass rod, armor-piercing bullet

lathe and cut a shallow cavity into its base. Though this gives a slight velocity increase, it isn't absolutely necessary.

Brass or copper armor-piercing bullets work best in barrels with a fast twist to stabilize the projectiles. With the .223, the 1-in-12 twist will greatly decrease the bullet's barrier-penetrating ability, while the new 1-in-7 twist will give good results.

A pistol bullet won't buck as much material as a rifle projectile but will defeat the ballistic armor. Brass pistol bullets also travel at a very high speed, so they have much the same effect as the French Arcane bullet and will create massive wounds at close range. In effect, the bullet's high speed makes up for its lack of expansion; it takes on the lethal high-velocity characteristics of rifle bullets. Because of this, it makes sense to keep pistol bullets small and light so that they can pick up all possible extra velocity. Remember, however, that velocity will drop off very quickly with distance; the bullet won't be dependable past 50 yards, though it will still be dangerous. You'll need to experiment to see what your lightweight pistol bullet is capable of.

Steel armor-piercing bullets offer a bit more penetration than brass or copper, but are a lot harder to make. The steel itself can't touch the barrel or it will cause excessive wear or may become lodged in the bore. The solution is to make the steel bullet's diameter smaller than the bore, then use a copper jacket to seal the bullet against the rifling. This is a little tricky, since the jacket can't be left behind in the bore after the steel bullet exits.

Turn the steel bullet in the same way as the

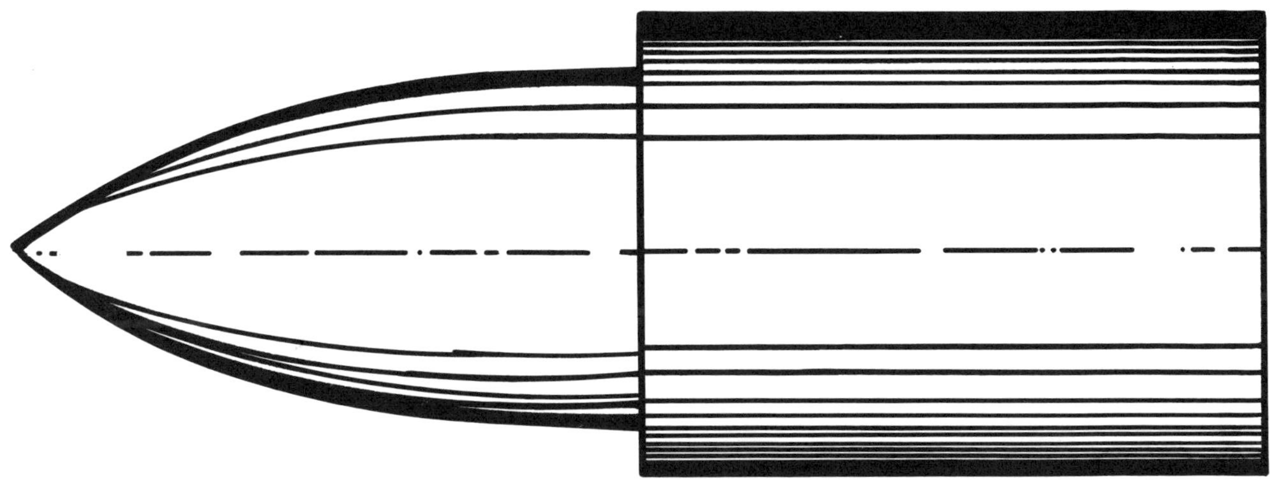

Steel rifle bullet with a short copper jacket.

brass, but make its diameter about 0.08 inch less than normal. Relief grooves won't be needed. When the overall diameter is within safe limits, turn down the base of the bullet so that it can be fitted into a copper jacket, and cut it off the rod.

The best way to create a copper jacket is to cut the top off a standard flat-based, hollow-point bullet and ream it out (again, the lathe or drill press will make this task easier). The steel rod can then be press-fitted into the jacketed base (a very tight fit is essential). If you have swaging equipment, it is possible to swage a bullet jacket around the steel insert. Take care not to damage the die while swaging.

New armor-piercing bullets like the 5.56mm NATO use a steel insert to allow the bullet to defeat metal. The velocity of the bullet is thus maintained over long distances because most of it is made of lead. This creates a dual-purpose bullet that doesn't require rezeroing or other changes.

Flat-based bullets seem to give better results than boat-tails with this alteration and travel though wood in a straighter path. A do-it-yourself version can be created by drilling a small hole in the end of a hollow-point bullet and inserting a small sharpened piece of steel. Provided the fit is very tight, the lead of the bullet will hold the insert in place. As with other armor-piercing rounds, a lathe will greatly aid your work. (Other materials besides steel can be used, but unless you have a tungsten or depleted uranium rod around, steel usually works best.)

Another way to create such a bullet for larger calibers is to drill the hole and then screw in a small machine screw. Once in place, the screw head can be ground off and shaped into a point. For best results, use a steel screw; stainless steel is even better. Try to make these bullets a consistent weight by grinding off some of the steel on those that are too heavy. You can create the steel-insert-tip bullets in loaded ammunition but you *must* use only hand tools when doing so. Power tools could heat up the round or otherwise set it off, showering you with jagged bits of brass. A steel insert can also be used on a turned brass bullet, although this is harder to do since the two pieces have to fit tightly together; they can be soldered, however.

You can load any of these armor-piercing rounds using the reloading data in a manual. *Do not* use maximum loads or those shown elsewhere for combat rounds until you're sure they won't create excessive pressures. Remember that the bullets can penetrate building materials extensively and will require more material to stop them during practice. Though it is presently legal to make and own armor-piercing ammunition, this may change in the near future. Be aware of the laws before you create any of these bullets.

LIQUID-FILLED BULLETS

One of the deadliest rounds can be created by placing a small amount of liquid or semiliquid material inside the front end of a bullet. This will cause the bullet to expand very rapidly when it hits a target. The flip side of this is that the bullet will offer almost no penetration. Such bullets can be of use, however, inside a house, since a missed shot would be less apt to hurt the wrong person. Liquid-filled pellets also are used in police sniper work when it is necessary to instantly immobilize a hostage-taker or the like.

Mercury is sometimes used for the liquid filler in bullets. While the density of mercury makes it very desirable, it has several serious drawbacks: it is extremely toxic and thus hazardous to work with during the loading procedure, it solidifies below freezing and loses its liquidity, and it will contaminate the area it is used in and any wounds it might enter. Heavy grease, oil of various types, wax, etc., will give nearly as good results without these disadvantages. Richard B. McSwain, the first person I know of to actually write down how to make liquid-filled bullets, swears that bearing grease is the next runner-up to mercury. McSwain also developed the following technique for getting the material into the bullet. It works well, and I've never found any need to modify it:

1. Drill a hole into the nose of a bullet. (He recommends a 9/64 drill used to a depth of 3/4 inch for .30-caliber bullets and a 13/64 to 1/2 inch for .38 or 9mm bullets. Guesstimate from there.) It may be necessary to file a flat point onto spitzer bullets before you drill them.
2. Melt the grease or wax that you're using for a filler and pour it into the hole. Using the filler in its liquid form eliminates any voids in the depression. A large spoon over a flame will work for the small quantities of material you are working with.
3. Allow the filler to cool. If it shrinks slightly, this is not a problem.
4. Use a piece of buckshot just a bit larger than the hole (#4 buckshot for the 13/64 or #2 shot for the 9/64)to cover it. Seat the shot with

your finger. Ideally, a little grease will ooze out around the shot.

5. Use a 100-watt soldering iron to solder the shot carefully to the top of the bullet. No flux or solder is necessary since grease or wax will flux the lead of the shot and bullet. Be sure the seal is complete so that the grease cannot escape.

6. Reweigh the bullet and use a powder charge appropriate for its new weight.

If you have to use mercury—and I strongly advise against it—do not use a soldering iron to seal the bullet unless you have access to a lab hood exhaust system. Mercury will vaporize when heated, and its fumes are deadly. Don't take practice shots with mercury bullets; you'll poison the area where you shoot. Mercury is not something to play with and is best avoided!

If you have swaging equipment, it is also possible to swage a liquid-filled bullet. You'll probably want to use a lead base inside a copper jacket and then fill it with grease or wax. When sealing it, you'll need as small a point as you can create on the bullet. Finally, you'll need to solder the opening shut. These bullets have a tendency to splatter rather than penetrate, so be sure to test them out to be sure they are effective.

MODIFIED FMJ BULLETS

Full-metal jacketed bullets have been required on the world's battlefields since around World War I. The conventions of war ban all soft points, hollow points, or pure lead bullets, as well as poisoned bullets. For some reason, nearly all countries have abided by these rules while at the same time using flamethrowers, napalm, poison gas, nuclear weapons, etc.

Probably shortly after the first FMJ bullets were issued, some doughboy decided to tip the odds in his favor by carving off the metal ends of his bullets. This creates an expanding bullet, and there are records of this being done in both world wars; it is probably more widespread than any country will admit.

The point of the jacket should only just be removed. FMJ bullets don't really have a full-metal jacket; the lead is exposed at the base of the bullet. If you remove too much of the point, the bullet may shed its jacket—and leave it in the barrel. Care should be taken to keep the bore clean for the same reason. With FMJ pistol bullets, it is possible to drill a small hole into the point of a bullet; again, don't overdo things or the bullet may lose its jacket in the barrel.

Swaging presses and good swaging dies can be used to reverse an FMJ so that the base becomes an exposed-lead tip. These bullets work well but are a lot of work to create and are more suited to the hobbyist interested in saving money than the combatant.

Some rounds don't need to be altered. The new 5.56mm NATO round is probably better left unaltered, and the same is true of any bullet with an armor-piercing insert if you need to defeat armor.

TUNNEL BULLETS

These bullets, which have a lot of different names, date back to the old English "manstopper" revolver ammunition used by British empire forces to keep the natives under control, and also helped lead to the conventions limiting combatants to FMJ.

The bullet is usually all lead and has a massive hole in its front which can cut a caliber-sized hole in flesh or even open and flatten somewhat, though this may not have occurred in the low-velocity weapons in which the bullet was used.

Modern high-velocity ammunition has made mushrooming bullets possible, so that there is now no necessity for such large holes in the bullets. Too, high-velocity lead bullets tend to coat the inside of the barrel with lead. A heavily-leaded barrel can cause a tunnel bullet to shed a large chunk of metal and block the bore. All in all, a lot can go wrong with these rounds.

The idea of drilling a larger hole into a jacketed hollow-point is sometimes proposed as a defense round. In testing, these bullets are far from ideal, however, because they usually mushroom quickly and then lose either a ring of lead, all or most of their jacket, or both. This may limit the spreading of the lead and actually defeat the round's purpose.

If you need a round that mushrooms quickly, then a liquid-filled or carefully designed commercial bullet like the Hydra-Shok does a more consistent job. The "manstopper" was a good bullet in its day; today, modern high-speed rounds do a much better job. You might consider using such a bullet in an old .38 Special which does not propel its bullets at a high enough speed to expand, but what you really need is a new pistol!

SHOT AND FLECHETTE ROUNDS

These rounds are two good ideas gone astray, as far as I'm concerned. They take different shapes, and include small, pistol-caliber cartridges containing shot, long brass pistol cartridges crimped shut like a mini-shotgun shell, short pieces of wire in a shot cup, and so on. They are very lethal within a range of five feet.

That a load of shot can be more destructive than a single round is certainly proven almost every time a shotgun is fired, but a shotgun delivers sufficient power to penetrate far enough to be lethal. A shotgun sends out a huge swarm of shot, so that some will follow others into a wound channel. The energy of much of the shot is dumped on top of a wound created by the first pellet, so that the damage piles up, creating a massive wound at close range.

A pistol has a rifled barrel, however, so that the shot spreads out excessively, and a handgun is limited in how much it can throw. Thus, the shot tends to create individual wounds rather than piling multiple impacts into a small area. While the shotgun creates a massive rat-hole wound at close range, the pistol makes a splattering of tiny wounds with little penetration. Unless one of the pellets hits an eye or is fired within about a four-foot range, the shot load won't have much effect on a determined attacker. If he is wearing heavy clothing or a ballistic vest, forget it.

These loads might be of use to someone living in an ultra-small apartment with paper-thin walls, but there are much better rounds for this and other combat situations. The idea is sound, but the weapon too small.

Like other ideas that sound good in theory but aren't too great in practice, the idea of shot loads for defense continues to surface from time to time. Don't waste your time with these variations. The shot load isn't that reliable and depends on delivering a lucky shot at extremely close range.

One variation on this idea that does work fairly well has been created by gun writer Joe Zambone. His "junkyard dogs," as he calls them, are created by placing various types of shot and wire in CCI/Speer shot capsules. Zambone uses .44 Magnum shot cups and cuts down the lids to create the maximum shot capacity. The best results are obtained with two sizes of brass brazing rod. Small links are made from 1/16-inch rods and larger diameter ones from 3/32-inch rods. The links are the length of the inside of the shot cup. Eleven of the larger-diameter rods will fit into the cup with three smaller ones nestled in with them.

The cup on Zambone's loads is placed over 7.0 grains of Unique in Federal .44 Magnum brass with a CCI #350 Magnum primer (if you try this, start out below this load to ensure that it isn't too hot for your firearm). These loads give the large links about 33 foot-pounds of energy and the smaller links around 15 fpe. The total load has 394 fpe at the muzzle.

The load is quite effective at 5 feet and remains so out to 21 feet, where a target received 11 hits in a wounding zone. It would appear that within 20 feet this load can create a disabling, if not fatal, wound. The plus side is that it allows a .44 Magnum to be used with minimal wall penetration. It also allows you to score a hit even with a slight aiming error. On the negative side, the load is very short range, and one has to wonder about the legality of a round which may cause terrible superficial wounds rather than "simple" gunshot wounds. Personally, I'd prefer to use standard expanding bullets which give a larger range, though the idea of preventing harm to bystanders by overpenetration is a good one.

Shotshells are offered commercially by CCI. In addition to .22 LR and .22 WMR, they also have .38 Special and .44 Magnum shotshells. Shot capsules are offered by CCI under the Speer label in both these calibers for those who wish to reload their own shotshells or create Zambone's "junkyard dogs."

MULTIPLE-PROJECTILE LOADS

In theory, multiple hits increase lethality by the square of the number of hits. Thus, three hits in quick succession are nine times as bad as being wounded once. Another factor is that with a slight dispersion, the shooter is more apt to hit his target. Thus, a three-round burst which spreads a bit will compensate for aiming error and may give one or more hits where a single shot would have missed.

The catch is that three-round bursts eat up a lot of ammunition and shotgun shells are bulky. Too, until recently, many weapons had a full-auto rather than three-round burst mode which, in the heat of battle, often became a "30-round burst mode," making assault rifles into single-shot weapons for all practical purposes. Finally, a lot of combatants are restricted to semiauto weapons.

One way around these problems is to create a

pistol or rifle round with multiple projectiles in each cartridge. These have to be larger than shot or wire to overcome range limitations. Fortunately, we don't have to start from scratch; the military, several commercial companies, and a host of reloaders have already done the groundwork.

The U.S. military used some multiple projectile loads in Vietnam. These were created by nestling five equal-sized, hollow-based, cone-shaped bullets inside one another inside a .50-caliber machine gun cartridge. Each shot propelled five bullets out the muzzle. There was also a choke attachment that swaged the bullets down to .30 caliber as they left the barrel. While these apparently worked well, they never really caught on, and research has been all but discontinued.

Such an arrangement would certainly have some advantages with .308 or .30-06 rifles. Three-round bursts of small 55-grain bullets seem to be more effective in combat than one shot from a heavier rifle; the .308 could hold three 55-grain nestled bullets and create a "three-round burst" with each pull of the trigger. This would mean that a .30-caliber rifle with a 20-round magazine, would have firepower equal to a .223 rifle with a 60-round magazine.

The big problem is that there is no good way to make such .30-caliber bullets at this time. Probably swaging equipment would be the best route to take if you're really keen on this idea. The problem in making such bullets is that they need a very sharply indented base for the next bullet's point to nestle into. Accuracy would also suffer with these rounds but accuracy is probably not a prime consideration in a three-round burst.

Multiple-projectile pistol ammunition can also be created. The best bet is to place a smaller-than-normal bullet on top of flat discs the same diameter as the cartridge; usually two of these discs are placed under the bullet. When such a round is fired, the bullet travels to the point of aim while the two discs tumble due to air resistance and will impact below, and often to the side of, the bullet. This creates a little spread to allow a hit even if a small aiming error was made, and gives multiple hits when the aim is good and the range close enough that the discs' velocity doesn't drop off.

Another method with less effective range is to place five discs in a pile and forget the bullet. This creates a flock of tumbling projectiles that are very lethal at normal pistol ranges. You are, however, up the creek if a foe moves out of close range and continues to engage; the velocity drops off drastically beyond 25 yards.

A very simple way to create the bullet and disc combination is to cut through a lead pistol bullet at a couple of spots (lubrication rings make nice guides). This can create a couple of discs and the bullet, but lead bullets aren't too great for combat. Unless you're stuck with an old .38 Special that can't take +P loads, this won't be of much use.

A more reliable combat disc can be created by swaging a gas check (which is normally used as a base on some lead bullets) and filling it with lead. You'll have to do a little modifying on most swaging punches so that such a thin "bullet" can be made. It's worth the effort, though, because once the modifications are made, you can turn out the discs quite easily. If possible, you should also make the nose of the bullet to fit into the base of the copper gas check. The best way of doing this is with a modified nose punch in the swaging die, but you can also get a jam fit by putting however many discs you use into a normal swaging die and scrunching them together.

If you wish to place a bullet over two discs, use a smaller than normal bullet to make room for the discs. Weigh the entire bullet-disc group and use reloading data for jacketed bullets. Don't try maximum loads until you're sure everything is working correctly.

The disc load is most suited to the .38 Special or .357 Magnum but will work with the 9mm Luger if extreme care is taken not to seat the discs too deeply. (The .357-caliber disc may otherwise bulge the case slightly or create excessive pressure.) Even after the round is perfected, don't use maximum loads with the 9mm. The .30 Carbine can also be loaded with discs made from .30-caliber gas checks thanks to its straight-walled case. The bullet/disc combination gives the rifle some long-range capability as well as "burst" effect for close combat.

It is essential that the discs be tightly seated so that they can't be pushed back into the chamber by recoil or chambering forces. Powder loads which nearly fill the case are a good safety measure for keeping the discs from dropping into the cartridge.

A sort of poor-man's multiple-projectile load is sometimes created by mounting several buckshot balls inside a pistol cartridge. The idea looks great at first; 00 Buck is .34-inch in diameter while 000 is .36, close enough to make them tempting in the 9mm and just right for the .38 Special or .357. With the straight-walled .30 Carbine case, No. 1 Buckshot is a handy .30-inch in diameter. It's simple to weigh out three pieces of shot, add the appropriate powder load, and seat the balls.

Unfortunately, I haven't found these rounds to be any good for combat compared to high-velocity bullets. Being lead, the buckshot can't be fired at any great velocity. While they do fit into the cases and barrels fairly well, they are hard to keep in place; they drift during recoil or chambering and raise pressure to a dangerous level.

In addition, a ball is one of the poorest ballistic shapes a bullet can have. Velocity drops quickly, so that the lethal range of these loads is quite limited. I'd suggest you avoid even trying the multiple-buckshot load.

SAFETY SLUGS

One shot load which doesn't work like a shot load but is highly effective in a pistol is the Glaser Safety Slug. If you've ever looked at Safety Slug prices and gotten a nosebleed from the high altitude, you're not alone.

One way out of this dilemma is to make your own safety slugs by swaging. The process is nearly as simple as the standard method of making a jacketed bullet; instead of using lead wire for the core, you use shotgun shot. The only tricks are keeping the shot from running out, keeping the weight standard, and sealing the nose.

Use the longest jackets you can find; otherwise, the bullets will be too lightweight and you'll have a hard time finding reloading tables for the projectiles. The shot leaves a lot of empty places between each pellet so that the finished bullet is rather light for its size.

Experiment with your swaging die adjustment depth, the depth of the nose punch, and different ways of sealing up the point of the bullet. The bullet itself is more apt to open up with a hollow point or breakable plastic cover, but you may also want to experiment with a flat point in the swaging die. (Doing so generally means buying another bullet ejector since it usually determines the point's shape—especially if you also use the die to make standard hollow-point bullets. You can turn an ejector out of a steel nail by using a power drill as a lathe. A file can be used to cut the nail to size and calipers show when to quit. Use the original ejector as a pattern for size measurements.)

To seal the shot in the jacket, use a hollow-point ejector in the swaging die. The pressure of swaging will wedge the shot into place (though a dab of epoxy glue might be a good safety measure).

Another way to seal the shot in the jacket is with a small square of metal cut from a *very thin* gauge of aluminum. Make the square big enough to cover the hole if you bend the four corners of the square down to fit into the jacket. When you swage the bullet, the tips will guide the square cover under the edge of the jacket and seal the shot in. You can also seal the shot in the jacket by creating a plastic "cap" by placing epoxy over the shot once the jacket is swaged into shape, or by using *very thin* lead wire (in the caliber normally used for swaging standard bullets) which is placed over the shot before swaging. Both of these methods can work, but *be sure to test either of these out.* If the jacket or the nose are too thick, the bullet won't open up.

If you have trouble getting the jacket to open, try using wire cutters to place eight scores or crimps at 90 degrees to the edge of the jacket before you load it with shot. The crimps must be made in several steps because the edge of the jacket will stretch and bell out. To get it back into shape, run the mouth part-way up into the swaging die.

Use the smallest shot you can find. Glaser Safety Slugs are generally loaded with #12 shot, which is the smallest available commercially. Larger shot creates more voids between pieces so that the round will have poorer accuracy and less weight for its size (resulting in higher initial velocity but poor long-range velocity and accuracy). Though you can buy bags of shot at many gun shops, it may be easier to buy some shotgun shells, peel open the front of the crimp, and use the shot in the shell. Since little shot will be needed for each bullet, this method is often cheaper than having to buy a huge bag of shot.

Although Glaser Safety Slugs used to be packed in liquid Teflon, I think this liquid was intended to stabilize the bullet rather than to create extra penetration. A cheaper alternative is to use oil inside the do-it-yourself slug and then very carefully solder the whole thing shut so the oil can't leak. Be sure to keep the bullet weight consistent from one slug to the next. This is a lot of extra bother, and I usually leave the liquid out and put up with the lack of pinpoint accuracy.

Make your first slug by capacity; you'll have to play around with the die adjustment and the amount of shot. After that, use a scale to weigh the shot, jacket, and front cover before you start; this will keep all the bullets the same weight when you're finished. Use the reloading data for a jack-

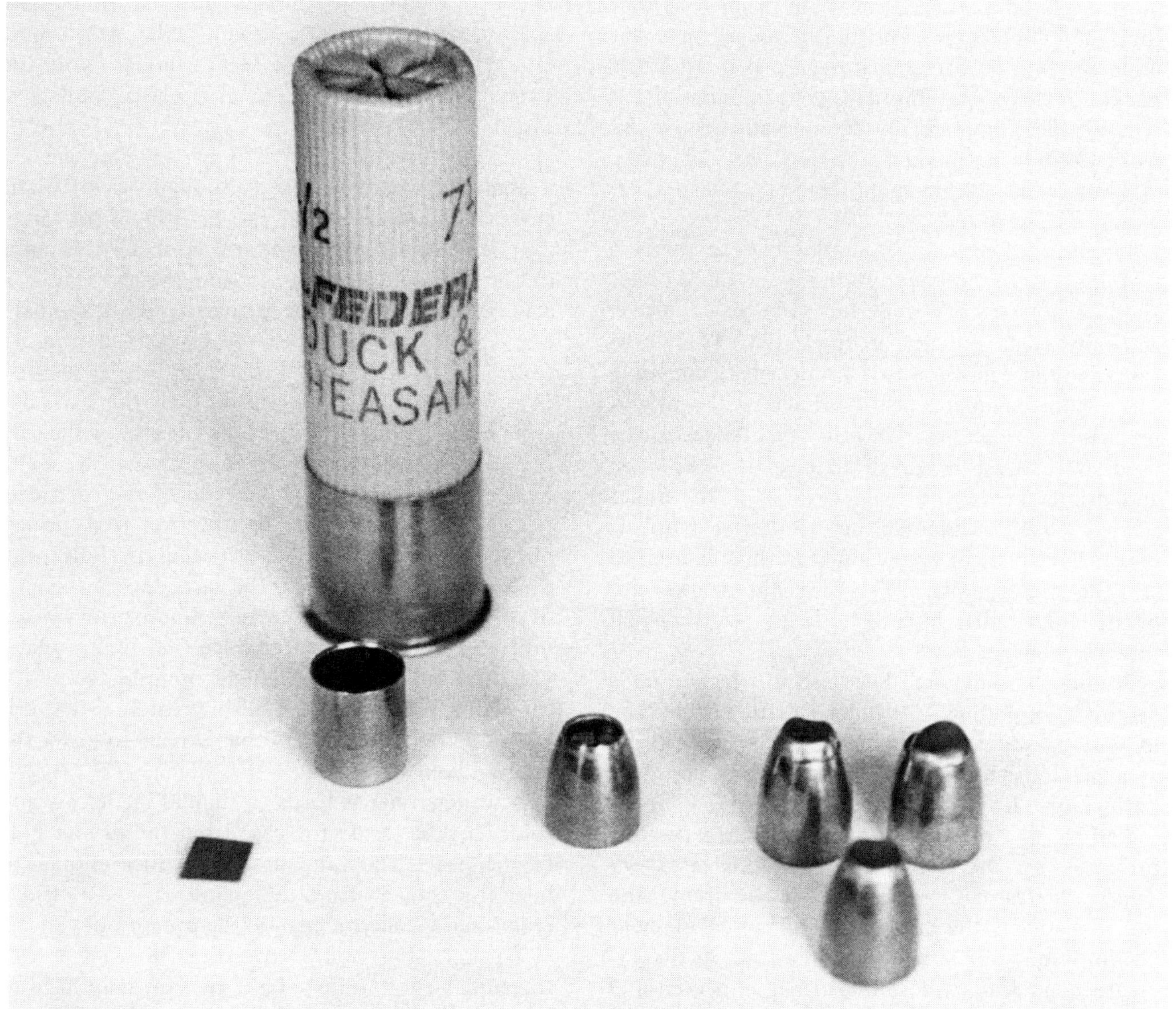

The "poor man's" safety slug can be created by swaging copper jackets filled with lead shot (obtained from the shotgun shell shown) and can be kept in place with a small square cut from an aluminum can.

eted bullet of the same weight, and don't use maximum loads until you're sure the round will function well.

As with all other rounds, you should carefully test the accuracy, reliability, and capability of your homemade safety slug. Start with the wet-newspaper test, check the brass of the first few rounds, and check accuracy.

A number of variations can be made on this load. One is to use clay, thick grease, or other materials to fill the jacket. The problem with using lighter materials is that the bullet is super-light and doesn't maintain velocity. One solution is to seat a small lead core at the base of the jacket for weight, using the other material to top off the bullet. The bullet will need to be soldered shut to prevent leaks in hot weather. If you try this type of bullet, test it for penetration and to be sure it opens up; if it just splatters on the surface of an opponent, you'll not be in an enviable position. Likewise, if it fails to open up, you've done a lot of work for nothing.

The safety slug is not normally used in rifles because it limits long-range accuracy. It might be of use, however, in "pistols" which use a rifle round like the Bushmaster or Enforcer or in .308 battle rifles with greatly shortened barrels. A safety slug would limit the dangers of overpenetration when the rounds are fired indoors, though such use is still very risky.

EXPLODING BULLETS

These sound good but are not a lot better than

most expanding bullets. The main problem is that they don't perform consistently unless they hit bone and will go off prematurely if they hit a belt buckle or heavy clothing. This type of ammunition is available commercially from National in .32 ACP, .380 ACP, 9mm, .38 Special, .357 Magnum, .45 Auto, and .44 Magnum. Dealer cost is approximately $1 per round.

If you need to make exploding bullets, the general idea is to hollow out the nose of a bullet (hollow points work well) and place an explosive, generally black powder, in the cavity. The whole thing is then capped with a primer charge of some sort. While standard pistol primers will work, a percussion primer or even a toy pistol cap can be used. Care must be taken to keep the explosive and primer below the protective rim of the bullet so that it isn't accidentally set off during chambering. A light coat of epoxy glue can then be used to seal the primer in place. This procedure works with pistol bullets; rifle bullets, however, are too small to work well.

About the only time I've used this technique is to create exploding .22-caliber air rifle pellets. The explosions add a little interest when you're firing at a metal backstop. Otherwise, I think you'd be better off to leave exploding bullets alone.

Perhaps the ultimate explosive bullet is the one proposed by Dr. Herman Kahn in his book *On Thermonuclear War.* He writes of a heavy rifle bullet made of the man-made radioactive element Californium 252. Such a bullet would, in theory, create a nuclear explosion when it impacted and compressed. The explosion would be small as nuclear blasts go—only equal to several tons of TNT! The main problem with such a bullet would be handling it and firing it at a high enough angle to keep the shooter out of harm's way.

PRACTICE LOADS

It's nice to be able to go off to a field or range and blaze away with your favorite firearm. Practice is important, but sometimes it's a real pain to load up the car and drive to the range. You spend more time getting ready and on the road than you do shooting. Wouldn't it be nice to have an indoor range in your basement so that you could just walk downstairs and spend all your time shooting?

The answer is to use your own weapon with greatly reduced loads. One way of doing this is to make a solid brass "cartridge" with a primer in one end and a pellet of some type in the other. With .223 weapons, the .22 air-gun pellet works well. Other calibers are a problem, though you can swage small lead bullets or cast plastic bullets in pistol dies.

With pistols and the .30 Carbine, you can use a standard cartridge case, although Speer plastic cases and bullets, which can be primed for target practice indoors, are more satisfactory. These come in .38 and .44 caliber. A .45-caliber plastic bullet is also available from Speer for use with standard brass in the .45 Auto, .45 Auto-rimmed, and .45 Colt.

With rifles, you'll need to turn the cartridge from a brass rod and drill a hole the size of the caliber in its forward end just deep enough to seat a bullet. From the head stamp end of the cartridge, you'll need to drill a slightly oversized primer pocket and then drill an oversized fire hole from the pocket, clear through the cartridge, to the base of the bullet pocket. File an extractor rim on the unit and you've got a cartridge that will propel a practice bullet with enough oomph to punch through a paper target. The only catch is that it is a single shot and you'll probably need to lower the zero on your firearm.

Another route is to use a standard rifle case and solder a tube inside the case from the primer hole to the neck. The tube should be thick enough to keep the "bullet" from falling into it. The cartridge is finished when you enlarge the primer pocket.

To load the unit into a firearm, first "seat" the bullet (if it isn't a tight fit you may have to load the unit into the chamber with the bullet pointing slightly upward), then point the barrel down, drop the primer into place, and slowly close the action. With some rifles, this isn't too hard, especially if the bolt can be held open; with others you need an extra hand or two.

It's possible to make a tight primer pocket and bullet fit so that the round can be cycled through the magazine, but this creates problems when it's time to reprime the cartridge. You have to take it to a press and a special tool will be needed to deprime it. You can't win with this design unless you want to spend a little money. Harry Owen comes to the rescue with his cartridge adapters. With the .223, a brass insert nearly identical to the one outlined above is available for $18.50 with a depriming rod costing $4.75. This unit allows you to fire .22 air rifle pellets without much problem. A Harry Owen cartridge adapter for the .22 rimfire comes in .308, .30-06, and .223. This allows you to fire

.22 CB Caps as well as standard .22 ammunition in your rifle. While the standard .22 is a little too much for indoor use, the .22 CB Long Cap available from CCI has only a little more power than a good air rifle (which can still be dangerous—use either projectile with care and be sure you have an adequate backstop). The .22 cartridge fits into the steel adapter, and an insert goes behind it. The insert is held in place by an "O" ring and transfers the energy of the rifle's firing pin to the edge of the rimfire. The .22 bullet hugs the rifling very well on a .223 rifle. The .30-caliber cartridge adapter contains a short rifled barrel; velocity isn't as good, but it works better than one might think. The adapters can also be used with standard ammunition if you wish to use the rifle for small game or quiet shots outdoors.

The Harry Owen .22 adapters cost $35 for the .30-06 or .308; the .223, which doesn't need the integral barrel, costs $18.95 for a steel unit or $27 for stainless steel.

The .22 CB Long Cap can also be used in .22 adapters for combat firearms. If you have one of the units, the CBs will have to be hand-cycled after each shot but will allow you to spend more time shooting rather than reloading adapter cartridges.

With revolvers, it is possible to fire wax bullets. These can be made by enlarging the pocket to take a larger primer (shotgun primers work especially well) and using the cartridge like a cookie cutter to fill its mouth with a load of wax. Paraffin works well; just pour it about a half-inch deep in a pan and load the brass before it hardens completely. The nice thing about wax bullets is that since many can be made at once, time isn't taken up punching out primers to reload for more practice.

I *do not* recommend that you use wax bullets for mock gun battles. The wax bullets can easily put out an eye and will leave a nasty bruise at close range, and there is no provision to keep the pistol from firing if a live round should get mixed into the practice ammunition.

Whatever type of brass you use for practice, be sure to mark it so that you don't try to reload it later on. While most of the adapters can't be reloaded, the wax-bullet pistol rounds can. This could create some dangerous head-spacing problems since the firing pin banging against a cartridge that doesn't actually fire will shorten the case quite a bit before you load it up with powder. Art supply stores often carry a chemical, usually a sulfate compound, that can be used to put a permanent black oxide over the practice cartridges so they can be recognized at a glance.

Though you'll have to use them at the range, because they're full powered, both Glaser and Super-Vel manufacture less expensive editions of their ammunition for practice. These rounds don't perform well at all as combat ammunition, but they do give you an idea of how the actual combat rounds will perform in your firearm. If you've decided to use either of these manufacturers' ammunition in combat, then it would be wise to work out with their practice ammunition.

If you use a regular range for practice, the new plastic-cased pistol ammunition from the United States Ammunition Company may be of interest. Currently available only for the .38 Special, the rounds have a plastic case that can be reloaded by press-fitting a specially designed bullet into the case after it has been reprimed and filled with powder. When the plastic shells get stretched out of shape, they can be boiled and will "remember" their original size. If you don't want the expense of reloading equipment but would like to save some money, the reloadable plastic cases are the thing to use.

Federal Nyclad ammunition has also found favor with shooters who want an inexpensive full-power practice load. The lead bullets in the Nyclad ammunition are coated with nylon, which reduces barrel leading to almost nothing and also reduces lead contamination in indoor ranges.

SILENCER ROUNDS

Silencers (sound suppressors to the purists) don't make the soft sounds of the movies unless you have a super-long custom silencer mounted as an integral part of a .22 rifle. All large-caliber suppressors operate with a lot of racket, nearly as much as an unsilenced .22 short in larger pistol calibers and all centerfire rifles. But silencers do change the sound envelope enough that many people mistake it for hammering or engine noise. The sound goes more or less unnoticed if the shooter isn't in sight, provided the bullet travels below the speed of sound. If the bullet travels above the speed of sound, it creates a sonic crack nearly as loud as the muzzle blast. While it is slightly different when a silencer is used and hard to locate, it will still be recognized as a gunshot.

There are three solutions to this problem, one of which is to slow the supersonic bullet down in the barrel/silencer so that it exits at a slower speed. This method allows the use of standard ammunition, but the bullet lacks power. Another solution is to use heavier bullets that travel below the speed of sound. Finally, you can choose a caliber that normally travels below sonic speeds, i.e., below roughly 1130 fps. Standard .22s and the .45 ACP are in this category as are other pistol rounds loaded with larger bullets and light powder charges. Revolvers don't work with silencers because too much gas leaks out between the cylinder and the barrel, and autoloaders create a lot of noise when their action cycles.

Suppressors on rifles work but not all that well. When you load up a bullet heavy enough to be subsonic, the zero goes out of whack and the normally flat combat-rifle ballistic arc looks more like a short rainbow. Consequently, snipers using a rifle often fire normal loads from very well-concealed positions. The bullet creates a loud sonic crack but the rifleman's position is not given away by the muzzle blast which, with an unsilenced weapon, normally reaches the area the bullet traveled through a bit after the supersonic crack takes place. Because the bullet is supersonic, it gets "there" before the sound of the rifle discharge when a silencer isn't being used. A skilled combatant can pick out a shooter's location by drawing a mental line between him and where the sound of the report comes from. By counting the time between the crack of the bullet and the thumping of the report, he can even tell how far away the shooter is. The military calls this the "crack and thump" way of locating a concealed sniper. When the crack is heard and the silencer cuts down the report (which can be heard from over 50 to 100 yards), it's quite hard to locate the sniper; those fired at know someone's shooting, but don't know from where.

With automatic pistols or submachine guns things are a bit better. Since silencers are generally designed to be used to surprise a foe, the shot can be taken when a carefully aimed hit in a vital area is assured, and the enemy is not charged up with adrenaline and can be stopped with a lighter, less powerful subsonic round. Since shot placement and surprise are the keys to lethality with a sound suppressor, small calibers like the .22 LR, .32 ACP, or .380 Auto may also be used for close-range silencer work.

With these many possibilities and limitations in mind, there is a fairly wide range of commercially loaded subsonic ammunition which will give fairly good results. Most .22 LR ammunition, except for "high-" or "hyper-" velocity loads, is subsonic. Nevertheless, Geco markets subsonic .22 LR ammunition designed for use with suppressors. National offers a subsonic 9mm Luger round at nearly the same price as standard ammunition while American Ballistics Co. offers subsonic loads in a number of calibers.

If you reload the 9mm and .45 Auto, it is possible to create ammunition for use with a silencer. Some subsonic loads for these calibers are shown in the "suppressor loads" table.

SUPPRESSOR LOADS

Bullet (grains)	Powder (grains)	Velocity (fps)	Energy (fpe)
9mm Luger-130 (jacketed)	Bullseye (3.8)	900	234
9mm Luger-130 (jacketed)	Unique (4.6)	900	234
9mm Luger-130 (jacketed)	Herco (4.6)	900	234
9mm Luger-130 (jacketed)	HS-6 (6.9)	900	234
9mm Luger-130 (jacketed)	Blue Dot (6.3)	900	234
9mm Luger 158/160 (lead)	Bullseye (3.0)	900	286
9mm Luger-158/160 (lead)	Unique (3.7)	900	286
9mm Luger-158/160 (lead)	HS-6 (4.4)	900	286
.45 ACP-200 (jacketed)	Red Dot (5.4)	950	401
.45 ACP-200 (jacketed)	HS-6 (9.2)	950	401
.45 ACP-230 (jacketed)	HS-6 (9.2)	900	414
.45 ACP-230 (jacketed)	Unique (6.9)	850	369
.45 ACP-230 (jacketed)	Herco (7.7)	850	369

While police departments, the military, or other government agencies can get silencers without paying taxes or going through a lot of red tape, the same isn't true for the average citizen in the United States (in Europe, silencers are all but unregulated). Before you run out and buy a lot of subsonic ammunition and plunk down $200 tax for a suppressor, I'd suggest that you read J. David Truby's book *Silencers in the 1980s* (available from Paladin Press) for a good idea of what is available and what silencers can—and cannot—do.

RIFLE AND PISTOL GRENADES

Hand grenades are very limited in range, being "hand launched." In an effort to overcome this, a number of systems have been developed to launch grenades of various types by using the firearm as a type of makeshift mortar. Most of these methods have had only limited success. The recoil created in launching a grenade is hard on the shooter and even if the rifle butt is placed against the ground instead of the shooter's shoulder, a steady diet of grenades can ruin a firearm.

The Mark 2 "Mills Bomb" was developed for the M-1 Garand rifle during World War II. It mounts on the rifle barrel with a special adapter (the M7 launcher) and the rocketlike grenade is propelled with a special blank cartridge. A special sighting system is normally mounted on the rifle for maximum accuracy. The M14 rifle has a wealth of different rifle grenades available for it, including the M1A2 adapter which allows regular hand grenades to be mounted and fired as rifle grenades. Most of the grenades for this rifle are outdated. Any grenade use with the M14 should be done with the M1A2 adapter and new hand grenades or with the "universal" grenades which fit both the M14 and newer rifles like the M16/AR-15.

The very common "Energa" rifle grenade can be fired from any model of the M16/AR-15 except those with extra long or short barrels, as well as other rifles with the AR-15-style flash suppressor. The grenade is propelled from the rifle with a special blank cartridge to a maximum range of 350 meters.

All the rifle grenades mentioned so far have more or less fallen into disuse mainly because they are hard to aim, requiring a special range finder to be accurate at long range, and use a special blank cartridge, so the weapon is tied up until the grenade is launched or removed along with the blank cartridge.

Though a special FFL is necessary to purchase and own rifle grenades in the United States, it is legal to purchase rifle-launched parachute flares and practice grenades. Phoenix Systems has a good supply of these products along with the special blanks and a retainer spring to keep the grenade secured on the rifle.

France has developed rifle grenades which work on most modern assault rifles with the "squirrel cage" flash suppressor. These have a bullet trap in the base which makes it possible to fire them with regular combat ammunition. The French rifle grenades are thus much easier to use since they can be quickly placed on the firearm's barrel and fired or removed if normal use of the rifle is required.

The Luchaire Company currently manufactures the grenades. The company claims that their special range finder can place a grenade through a window-sized opening at 150 meters. Antiarmor rounds capable of defeating the side or rear of tank armor (250 to 300mm penetration of rolled homogenous armor), and antipersonnel grenades with 60mm armor penetration and a lethal radius of 11 meters are available to qualified purchasers. If you need a rifle-launched grenade, the French models are probably the best available.

Israel Product Research Company has produced a riot/crowd control grenade called the "Rubberhead Projectile M-809." Its payload is a large rubber ball designed to hit a troublemaker at 160 yards. It gives a knockout punch which is less apt to injure the person than plastic or rubber bullets. The projectile can also be filled with 80 grains of tear gas so that others in a crowd will be affected. Like earlier rifle grenades, the Rubberhead requires a special blank for firing.

Work has also been done on rockets which propel themselves after launching. These have less punishing recoil and can be aimed with the standard rifle sights since the projectile contains propellant to compensate for gravity. Such devices are an intriguing idea which offers a large increase in accuracy to the basic rifle grenade concept.

I *do not* recommend that reloaders try to create the ballistic rounds used to launch grenades. The chamber pressures needed to launch grenades are sometimes quite high, and any error in the loading would be disastrous.

* * * *

Special bullets can allow you to do things you can't with standard ammunition, but you always sacrifice some other ability to make any gains when it comes to ammunition. Don't be seduced into creating fancy loads that do tricks but may be useless in combat. You're better off with standard ammunition that will do the job rather than trick bullets that may—or may not—help you when your life is on the line.

8. Combat Rounds for Shotguns

A shotgun is an awkward firearm. It's large and uses huge cartridges. To date, no one seems to have marketed a semiauto action that can digest a wide variety of loads reliably. Combatants have to either put up with the excessive recoil of a pump action or with the many breakdowns that plague most modern semiautos.

Before you buy a shotgun for combat, you should read *The Combat Shotgun and Submachine Gun* by Chuck Taylor (available from Paladin Press). He gives a fair view of the two weapons and in the process gives a good idea of the combat shotgun's limitations.

If after all this you decide to carry a shotgun, there is a wide range of ammunition to choose from. While some is designed principally for sporting use, most all rounds can be used in combat situations. The 12-gauge shotgun is most often chosen by combat shooters. Shooters who find the kick of the 12 gauge excessive may wish to drop down to the 20 gauge. A quick look at the chart below shows that by using magnum loadings in the 20 gauge it is possible to equal many of the standard 12-gauge cartridges. The only problem is that the 20 gauge has less commercial ammunition to choose from.

Other gauges like the 10 and 16 would also be suitable for combat but usually aren't recommended because of the difficulty in finding ammunition and reloading components for them. If these aren't considerations for you, then you'll find that you have a number of gauges to choose from.

Here's how the different gauges stack up:

COMPARISON OF SHOT-SHELL GAUGES

Gauge	Length	Shot Weight (ounces)	Velocity (fps)	Energy (fpe)
10	2-7/8	1-5/8	1330	2793
10	3-1/2	2	1255	3061
12	2-3/4	1-1/8	1145	1433
12	3 (Mag)*	1-7/8	1210	2668
16	2-3/4	1	1220	1446
16	2-3/4 (Mag)	1-1/4	1260	1928
20	2-3/4	1	1165	1319
20	3 (Mag)	1-1/4	1185	1707

*Only 12-gauge shotguns chambered to receive the 3-inch magnum will work with this load. Most 12-gauge shotguns will not chamber the 3-inch shell.

Like pistol and rifle users, shotgunners should remember that ammunition manufacturers have pretty well perfected the art of making 100-percent reliable ammunition. Most handloaders don't have the ability or power of concentration needed to produce well-functioning, reliable loads time after time. You'd be better off using commercial ammunition for combat and relegating reloads to practice, for you can save a lot of money by reloading shotgun ammunition. If you use a pump action, you can save money and the aches and pains of recoil by using reduced loads for practice.

All is not lost in the creation of special combat loads for the shotgun, however, even if you can't reload consistently reliable ammunition. You can also purchase factory-made ammunition, crack it open, modify it slightly, and close things back up. Provided you use a little common sense and caution, you can create many useful types of ammunition with a bare minimum of reloading equipment without sacrificing reliability.

Whatever you do, remember that your life may one day depend on a reliable cartridge in the chamber of a shotgun. One failure to chamber, fire, or extract is all it takes to get you killed. Remember that what works in someone else's firearms won't necessarily work well in every weapon. Try things out before trouble hits.

Study the loads listed in this chapter to see what can be done and then, when reloading, back off the powder charges a bit when you try them out until you're sure that they won't be maximum loads in your own shotgun. Maximum loads vary according to a firearm's chamber space, temperature, bore size, choke, and so on. The shotgun seems to vary more in size and strength from one firearm to another than the rifle or pistol.

When you test a new load, always inspect the first few empty hulls for signs of excessive pressure. As with metallic cartridges, look at the primer to see if it is flattened or cratered more than usual; check for primers that are ruptured or leaked and for primer pockets that are overenlarged; and watch for hard extraction or failure to cycle in a semiauto action.

Unlike metallic cartridge reloading where you can try out a different primer or brass, shotgun reloading requires that you make *no* substitutions. Because of the shell size and rather fast powders that are used, a change can spell disaster. Try different things that are recommended by the reloading manuals but *never* try to create one of your own loads unless you have access to a laboratory equipped to check the pressures the cartridge creates.

If you reload, you should study only new reloading manuals. New powders and perhaps even cartridges will have been developed since this book was written. While the loads and ideas presented in this chapter work well, better ones are on the way, especially with the work being done on new combat auto shotguns by the U.S. military. Be sure you're not missing extra accuracy, range, or lethality with new shotgun shells.

As mentioned in Chapter 3, reloading is basically a matter of inspecting a cartridge to be sure it is still okay, depriming and resizing it, replacing the primer and powder, and placing the projectile(s). With shotgun shells, you also generally have to place a wad or wad column between the powder and shot and crimp the cartridge closed. It's all pretty simple and any reloading press you purchase will come with step-by-step instructions for its use.

It is important to keep your work area clean and clear of junk and have only the primers, powder, and shot you are using open on the reloading bench; otherwise, you may make a tragic error. Unlike pistol and rifle primers, shotgun primers can't be substituted even on standard loads. If a reloading manual calls for an X primer, don't try to substitute a Y primer or you may have an explosive situation.

The same is true of empties. Brand X cartridges will produce different pressure curves than brand Y. Don't substitute until you've learned the ropes and know which is which. Follow the reloading directions carefully. Generally a manufacturer will use the same type of cartridge for the same type of loads with different-sized shot. Thus, a number 9 shot empty and a number 7-1/2 within the same type and brand can be used for the same reloads provided they originally held the same weight of shot and powder charges.

The major types of shotgun shells are paper base and plastic base. Both will probably have a plastic case unless you've found some ancient paper cartridges or some newly made brass ones. To find out what you have, look inside the cartridge. The paper base has a paper insert inside the base, while the plastic base has a plastic insert either in its base or as part of the case molded into the base. In general, the plastic base will have less strength so care must be taken when approaching maximum loads.

There are a couple more important points about reloading shotgun ammunition. One is that both the powder and shot meter poorly. Shot-charging bars or cups will throw varying weights from one shot size to another. This is because the voids increase with the shot size so that a charge of No. 2 shot will weigh less than the same size charge of a smaller shot.

With powder, there is a similar problem if you use a press with an integral powder container and measure. Many flaked shotgun powders tend to settle with vibration. As you work the charging handle of a press, the powder will settle in its charging bar so that the charge will be greater than it would be if you just pulled a charge to weigh it. Coupled with the shot-weight problem, it is very easy to get dangerously heavy powder and shot loads. The solution is always to throw and weigh the shot when you start out and to weigh the powder charge after cycling a shell through resizing, priming, etc. A good powder scale is one of the most important pieces of equipment in reloading shotgun shells, especially if you're approaching maximum limits. With smaller scales, you may have to divide the powder or shot being weighed. When you do this, be sure your arithmetic is accurate when you add up the fractional weights. It's also a good idea to weigh shotgun cartridges after you've reloaded them so that any which haven't received an adequate charge of powder or shot can be discovered.

With combat ammunition, it is essential that rounds chamber easily. Therefore, the hull must be carefully resized. Shotgun shells, like their all-brass brethren, wear out as they are reloaded and should always be inspected for imperfections and signs of excessive wear. Signs of excessive wear include separation between the brass head and the plastic case, tearing of the crimp mouth, splits in the case body, and crimps which start to spiral rather than form in a star pattern. When in doubt, throw it out. For combat, use new shells or ones that have only been reloaded once. Save the record-number-of-reloads shells for practice.

A shell will chamber easily if its front edge is nicely rounded; this occurs only when shot, wad, and powder fill the shell exactly. If you have extra space or too full a cartridge, the crimp won't be right and the shell may not chamber at the critical moment. Many reloaders who plan on using their cartridges in combat will cycle all newly loaded ammunition through their shotgun. This is a good way to check the loads, but always keep the shotgun pointed in a safe direction; you may discover a poorly loaded cartridge when it accidentally causes your firearm to discharge!

A problem which many reloaders have is the "blooper" shell. This is a load that doesn't fully ignite the powder and "bloops" the shot, wad, and powder a few yards out the barrel. There are several causes for bloopers. One is the use of a primer without a foil cover over its flash hole which lets powder settle into it and restricts the primer's fire so that all the powder in the shell isn't ignited. This creates only enough power to push the shot, wad, and unburned powder out of the tube. Always use primers with a foil or lacquer cover over the flash hole. Another cause of bloopers is the use of slow-burning powders. While these slow powders may work for practice, don't trust them for combat loads.

Other bloopers can occur when the wad is not pushed fully down over the powder. This creates a space which allows the powder to be pushed forward when the primer goes off rather than being fully ignited. To prevent this, take care when purchasing wads; different types of cartridges require different wad designs. Currently, the market is pretty well limited to two types of wads. One is cylindrical in shape and the other is tapered. While the tapered wad will fit all shells, it allows powder to get up around it so that a blooper will often result. The cylindrical wad inside a tapered shell may either compress and leave a space over the powder or cause the crimp to pop open if the shells are stored for long. A blooper can also take place when the crimp doesn't hold the shell together long enough upon ignition. This is just another good reason to discard empties with torn crimps.

Bloopers can be caused by foreign material in the shell, such as dirt, grease, and hair, that can get into a shell when it is on the ground. Burnt powder can foul the inside of the primer pocket. Failure to check the inside of the cartridge and the primer pocket during reloading is a good way to have a blooper.

Water can create bloopers by getting into a shell before or after reloading. Be sure shells are dry inside when you're reloading them and then strive to keep them dry. Shotgun shells are very leaky. Keep your shells away from moist conditions or they may fail. A ring of shellac or fingernail polish around the primer and a drop of wax over the crimp will help keep water out, although neither is 100-percent effective.

One of the major reloading expenses is shot. The best patterns will be produced by chilled shot, which has antimony in it to make it harder. "Drop shot" is nearly pure lead and will tend to deform when it is fired. This, in turn, will create some very poor shot patterns from time to time. Use chilled shot so that your foe won't luck out and be in a hole in your pattern when you need to zap him.

Also, stay clear of recycled shot except for plinking.

Whether you reload or are buying commercial ammunition, the matter of "drams" and "dram equivalent" can become a mind boggler. These terms are more than a little confusing, but continue to be used even though they have lost much of their meaning.

Powder loads in shotguns are measured by weight in grains just like other small arms ammunition. The catch is that shotguns as a firearm type are so old that the major calibers were originally loaded with black powder. Black powder was measured by volume in drams rather than weight. During the switch to smokeless powder, the energy of the loads became more important than the amount of smokeless powder. Thus, it was essential to let buyers know the equivalent of the smokeless load in black powder, and the "dram equivalent" was born. Now most seasoned shooters know what they want by this standard rather than by how much powder the shotgun actually has in it. Attempts have been made to change the system but sportsmen resist such logic and new shooters just have to learn to abide by it.

Another carry-over from the past is "high brass" and "low brass." This has to do with the brass head and rim on most shells. The marketing idea was that the larger "high brass" running up the side of the case would show users that the cartridge was stronger and could contain a larger powder load. In fact, the extra brass adds a little strength to the cartridge. This marketing gimmick has survived despite the fact that most shooters know that the brass has nothing to do with the shells' strength. On new shells, the brass may have something to do with the power of the load, but high or low brass should not be used to determine how much pressure a cartridge can take when reloading. A good example of how little the brass has to do with things is the all-plastic case which has a plastic, rather than brass, head and rim. It is every bit as tough as traditional shells.

Gauge and shot sizes suffer from much the same antique terminology. Gauge (of all but the more recent .410) is determined by the number of pure lead balls with a diameter equal to that of the bore equal a pound. For example, the 12 gauge, which has a bore diameter of .729, is measured with .729-inch lead balls. If you weighed these balls, twelve of them would equal a pound.

Shot size makes only a little more sense. It moves up in more or less steady increments, but numerical designations don't correlate to the steps other than the larger the number, the smaller the size. Sometimes a super-small shot is available (called "dust"), but generally shot sizes range from the smallest No. 12 shot (.05-inch) to No. 2 (.15), and then things get a bit complicated. The next two steps up in size after No. 2 are "Air Rifle" (.175), and then "BB" (.18). After this, we get into the buckshot group, which starts with No. 4 Buck (.24), then moves on to No. 3 Buck (.25), No. 1 Buck (.30), No. 0 Buck (.32), No. 00 Buck (.34), and No. 000 Buck (.36). The main thing to remember is that the larger the number, the smaller the shot size and that the more zeros, the bigger the shot. We should all be thankful they didn't use decimals when it came to naming buckshot.

Unfortunately, this backward-looking terminology and the lack of a new beginning when moving from black powder to smokeless has saddled the shotgun with two major problems that affect combat shooters:

1) The shotgun itself is not as strong as it could be. Given a stronger mechanism, the firearm could enjoy longer ranges and increased lethality with modern powders.

2) The shotgun shell is larger than it needs to be. Wads now have to fill up the area once occupied by powder. Anyone who has carried a shotgun into combat knows how nice it would be to have smaller shotgun ammunition.

CHECKING OUT A SHOTGUN

Not all shotguns pattern well. Your first step when checking out a shotgun should be to fire it at a few large cardboard boxes or other targets that will give you an idea how it patterns. You may find that your shotgun shoots off to one side, above, or below your aiming point, or that its barrel doesn't pattern well. These problems should be corrected by a gunsmith. You have enough worries in combat without having to remember to aim high and to the right all the time.

A 22-inch barrel is all that is needed on a shotgun to fully burn most powders. A longer barrel may look like it will give extra power and range, but it doesn't. It can make sighting easier and allow you to use a longer magazine extension, but it doesn't increase the power of a shot. Consequently, many owners of combat shotguns prefer shorter barrels that are less restrictive to close-quarters movement. Barrels shorter than 22 inches will exhibit increased muzzle blast and flash because some of the powder will burn outside the barrel.

Because the shotgun is often used indoors, it would be good to test its performance at short ranges. The first surprise for many people is that the shot doesn't spread out much. A shotgun will create large "rat hole" wounds at about 10 feet with a cylinder choke, or out around 30 feet with a heavy choke, but the pattern will only be several inches across, not the basket size seen in movies. Poetic license only works on the silver screen; don't expect your shotgun to do the impossible in combat.

The "choke" of a shotgun is the constriction at the muzzle which channels the shot into a tighter cluster as it leaves the barrel. Chokes vary from almost none in the cylinder choke to the maximum full choke. The more choke, the smaller the shot pattern. The various types of choke include—from least to greatest—the cylinder, 1 skeet, improved cylinder, 2 skeet, modified, improved modified, and full choke. Any of these can be used to fire a slug as well as shot, but the best slug results are obtained with barrels with only a small amount of choke.

This leads to two important points. First, with small shot, the pattern density must be fairly large to create the massive shotgun wound. Beyond a certain range, the shot spreads and, while it may inflict a number of wounds, it will not create the massive wound necessary to decide a fight. The second point is that despite what many shooters think, the shotgun does not allow much aiming error. When the pattern is large enough to allow for error, it is not dense enough to create lethal wounds, and the shot will have lost much of its power because of its poor longer-range ballistic behavior.

While most authorities recommend a barrel that gives a maximum spread, a good argument might be made for the maximum choke for anything other than buckshot. The best solution is probably to use a shotgun with a variable choke, although some variable chokes cannot fire slugs without suffering damage.

CHOOSING THE RIGHT SHOT

Once you've picked your weapon and choke, the next consideration is the type of shot to use. Buckshot loads will allow a greater spread without loss of power, but are limited by their small number of pellets.

Let's take a look at the spread you can expect with various types of choke (these figures will vary somewhat from gun to gun and according to the shot used).

SHOT SPREAD
(in inches)

Choke	2 Yds.	5 Yds.	10 Yds.	20 Yds.	30 Yds.	40 Yds.
Cylinder	5	10	17	32	44	58
Improved	4	8	14	26	38	51
Modified	3	6	11	21	32	45
Full	2	4	8	16	27	40

Another interesting table shows the percentage of shot that will remain within a 30-inch ring over various ranges as the pattern spreads; the more dense the pattern, the deadlier it is. Here's a comparison between various bores at various ranges (these are very approximate as the type of shot, shotgun, wind, etc., can change things radically).

SHOT DENSITY
(percent)

Choke	2 Yds.	5 Yds.	10 Yds.	20 Yds.	30 Yds.	40 Yds.
Cylinder	100	100	100	76	44	35
Improved	100	100	100	90	72	50
Modified	100	100	100	96	85	60
Full	100	100	100	100	90	70

Now, the catch in all this is that you have special considerations for indoor combat. Buckshot is best suited for outdoor use because it can go through a number of layers of indoor walling and inflict serious injury on bystanders. Pellets smaller than buckshot will still be lethal at close range but less effective if fired through interior walls. Buckshot gives more energy at longer ranges than small shot, but the number of pellets in a buckshot wound is limited so that at 25 to 30 yards, large holes become apparent in the pattern. With cylinder choke, the pattern may be large enough to lose a man-sized target at 30 yards. One solution is to use smaller shot but again it will drop its energy rapidly enough to severely limit the damage done by the pellets. While the perfect solution for long-range shots would seem to be a full choke, buckshot in fact tends to create doughnut patterns with a full choke—a large empty area in the center of the pattern at the point where you're aiming! This is why cylinder or improved cylinder chokes are generally found on modern combat shotguns.

There are two schools of thought about the best buckshot sizes for combat. One suggests that the shooter use No. 4 shot to maximize the chance of hitting the target, while the other suggests 00 buckshot for long-range effectiveness.

But in addition to the pattern spread, there is the problem of energy loss with round shot. For example, at 25 yards No. 4 buckshot will only penetrate 3 to 5 inches into ballistic gelatin. No. 1 buck does a little better with 6 to 8 inches; 00 buckshot, 7 to 9 inches. No. 000 buckshot gives overpenetration of 8 to 10-1/2 inches, so that it might exit the body of an opponent rather than dumping its energy. From an energy standpoint, the No. 4 buck is roughly equivalent to a .22 short bullet, 0 to a .22 LR, and 00 to a .32 ACP, while 000 is equivalent to a .38 Special! None of these single rounds is noted for its fight-stopping abilities, especially with a nonexpanding bullet, which is basically what buckshot is.

The shotgun is a close-quarters weapon. If you need to engage an opponent beyond 30 yards, you need something more potent than a shotgun loaded with buckshot. Given the limitations and spread of the various rounds, 00 buck may have a slight edge over the other loads, though it becomes a toss-up with variations of wound ballistics and range. All shotgun loads, however, have great limitations.

One solution to the shotgun's limits at extreme range is the slug. (Though these are sometimes called "rifled slugs," they are not. The shotgun has no rifling to grip the slug.) These heavy lead bullets are contained in cartridges with open ends, so that they are easily recognized. The shotgun fires these shells in the standard manner and the slug exits the barrel with a lot of energy, which it carries fairly well over 100 yards. However, the shotgun's smooth bore doesn't give the slug a lot of accuracy. An 8-inch group at 100 yards would be an excellent "bench rest" group for most shooters, while in combat shooting, hitting an opponent at 100 yards would be problematic. Even with slugs, the shotgun remains essentially a short-range weapon. But at close range, when the slug connects, the results are remarkably better than they are with shot. The various types of buckshot and slugs have the following amounts of velocity and energy:

MUZZLE VELOCITY/ENERGY FOR SHOTGUN PROJECTILES

	Weight (grains)	Velocity (fps)	Energy (fpe)
No. 4 Buckshot	20.5	1300	77
No. 00 Buckshot	54	1325	211
12-Gauge Slug	400	1300	1500
20-Gauge Slug	280	1300	1050

At extreme slug ranges, this table gives a realistic comparison since it is highly possible that only one buckshot pellet will hit an opponent. At closer ranges, more pieces of shot would connect; within 10 to 20 yards, depending on the choke, a full load of shot could connect. Thus, a more realistic close-range comparison would be by total loads of shot and buckshot as compared to slugs. In such a case, we obtain the following figures.

MUZZLE VELOCITY/ENERGY FOR SHOTGUN PROJECTILE LOADS

	Weight (grains)	Velocity (fps)	Energy (total fpe)
No. 9 Shot (710)	0.74 (each)	1200	1704
No. 4 Buckshot (27)	20.5 (each)	1300	2079
No. 00 Buckshot (9)	54	1325	1899
12-Gauge Slug	400	1300	1500
20-Gauge Slug	280	1300	1050

Coupled with the larger disabling factor created by multiple hits, it is easy to see why shot and buckshot loads are usually chosen over slugs for anything except long-range shotgun shooting or barrier penetration, which brings up the final consideration in selecting shotgun ammunition—penetration of barriers.

I once spoke to a policeman who had tried, along with several others, to stop a car with buckshot. Though they scored a number of direct hits on the car, some precisely placed on the driver's door, the buckshot didn't penetrate and the car made it through their roadblock. Buckshot, regardless of size, won't penetrate a car's body with reliability. In addition, when the shot hits sheet metal at an angle, it tends to ricochet; buckshot will penetrate side windows but is unreliable when fired at the windshield. Slugs generally will penetrate a car body, so if your opponent is behind a light barricade or inside a car, slugs might be the thing to use.

Slugs should be fired from a cylinder or improved cylinder barrel for best results. Barrels with rifle-style sights are also a must. Most double-barreled shotguns don't send a slug to the point of aim and are better avoided altogether.

INDOOR FIGHTING

What do we have thus far? First of all, the shotgun can be highly effective indoors. Because of the lethality of all types of loads at close range, the dangers of poor aim which might hit innocent persons elsewhere in a house can be minimized by using small shot. Size 12 would probably be best, though sizes 6 through 9 are easier to obtain commercially and would be stopped nearly as quickly.

Because of the close ranges involved in most indoor combat, there is no need to use magnum loads; standard hunting, field, or trap loads all have sufficient power and a large enough number of pellets to create a massive wound with one solid hit. They also allow quicker recovery between shots because of their reduced recoil.

Buckshot, because of its ability to penetrate walls, should be reserved for use when innocent bystanders will not be in or near the house. Buckshot enables a combatant to "stitch" walls and hit opponents behind them as well as engage enemies should the fight spill outdoors.

OUTDOOR COMBAT

The shotgun, given its present limitations, is a very poor outdoor combat weapon in any environment except perhaps the jungle. Even with slugs, the weapon has accuracy problems, and with shot it lacks both the accuracy and power to create lethal hits within the normal combat range of 300 yards or less. If the shotgun is deployed in units operating outdoors, it should be supported by riflemen.

When the shotgun's limitations are considered, it is little wonder that most combat troops are armed with rifles or machine guns. Yet, the shotgun has the potential for massive fire if anyone ever bothers to upgrade its primitive ammunition. Until then, don't forget the weapon's handicaps.

SPREADER LOADS

In close combat, extra spread can enable a shooter to score a hit even if his aim is off a little. This capability is not as great as many people think, however, as study of the above tables shows. Sometimes a little extra spread will help out, though, especially if you are using the weapon indoors at close range.

The spreader load is available commercially as the "Brush Load" or "Special Upland Load." They give an extra 25 to 30 percent of shot spread regardless of what type of choke the shotgun has. Spreader loads generally use special cards mixed in with the shot, though they are sometimes created with square shot. The card loads are probably a better idea since round shot retains its energy better.

These rounds are hard to obtain commercially, but they can be reloaded, or, if you want the reliability of commercial ammunition, you can open the shell crimp, change things, and then crimp it shut with one of the inexpensive Lee reloaders.

The simplest way to creat a spreader load is to bash the shot with a hammer so that most of it is no longer round. This amplifies the effect of air resistance and causes the shot to drift apart more rapidly, but, like square shot, it loses its velocity quickly. All in all, card inserts are the best way to enlarge the shot pattern.

There are two types of cards which can be inserted into the shot. Both can be created with lightweight shirt cardboard. (In a scrape, business cards or playing cards will also work.) The easiest method is to cut two cardboard rectangles whose width is equal to the inside of the cartridge and whose height is equal to the depth of the shot. At the middle of each rectangle, cut a slot halfway through, then slip the cards together in the form of an X. Place this in the cartridge so that the X can be seen when looking downward into the top. Fill in the shot around the cardboard and crimp the cartridge shut.

Another way to create a spreader load is to use the shot cup as a template to cut out two cardboard discs that will fit into it. Place one third of the shot into the cartridge, place a disc over it, set the next third of the shot over the disc, add the second disc, drop in the rest of the shot, and crimp the round closed. With loads using the spreader discs, you'll be able to add several inches to your

Buffered loads like this commercial one help to increase the lethal range of buckshot. Photo courtesy of Olin Corporation.

close-range pattern and decrease the danger to others from misses.

For maximum spread, the shot cup of the plastic wad should be removed. This creates some added space in the cartridge, however, and shouldn't be done unless you have time to carefully add shot or wadding to fill the space created by removing the cup. Removing the cup is not necessary to create spreader loads but will slightly increase the overall spread of the shot.

As with other reloads, check the crimp and exterior of the shell carefully to be sure the cartridge will work reliably.

BUFFERED SHOT

At the other extreme from spreader loads are buffered loads. Though the primary purpose of buffered loads in hunting is to keep the pattern more even, they also create smaller shot patterns with a higher velocity and energy level because the shot isn't deformed as much by the choke and has slightly better ballistics. As an example, a shotgun with an improved cylinder choke will only put half its load of buckshot in a thirty-inch circle at 40 yards. With a good buffered load, the same gun can place all of the buckshot in the circle, though it still does not keep the velocity high enough to be effective beyond 40 yards.

The easiest way to obtain buffered shot is to purchase commercial ammunition, but it's hard to find currently. Fortunately, it is also easy to reload buffered loads. A number of buffering materials have been used in the past with varying degrees of success. Among the worst are corn meal and flour which can explode with a nice fireball effect as the shot leaves the muzzle, and which can also attract moisture and create a gummy mess inside the shell. Bone meal, cream of wheat, and various types of soft plastic have also been used with safer, but somewhat unsatisfactory, results. The only reliable buffering material is granulated polyethylene, which can be purchased in most gun shops. Buffering material can greatly increase the pressure of a given load. Therefore, you must be careful; you *cannot* use standard reloading data for buffered loads.

Here are several loads that work provided you use the exact components shown:

12-Gauge Buffered Buckshot Loads

Shell: Peters' Blue Magic
Primer: Winchester 209
Powder: 31.0 grains Hodgdon HS-6
Wad: Remington SP-12
Shot: 12 #0 buckshot
Buffer: 22 grains of granulated polyethylene
Velocity: 1360 fps

Shell: Peters' Blue Magic
Primer: Winchester 209
Powder: 31.0 grains Hodgdon HS-6
Wad: Remington SP-12
Shot: 9 #00 buckshot
Buffer: 22 grains of granulated polyethylene
Velocity: 1360 fps

SILENT CARTRIDGES

Silencers are nearly worthless on shotguns. Because of the large volume of gas released when a shotgun is fired, a silencer would have to be huge to work. The U.S. military, however, developed a metal cartridge that fired nearly silently. This little-known shell was made with a flexible metal sleeve attached to the inside of the cartridge between the powder and the shot. The sleeve formed a cup which held the shot, and when the round was fired, the powder expanded and turned the sleeve inside out. The explosion of powder was contained inside the metal sleeve, while the shot was launched from the barrel by the sleeve's violent movement. Because the gas was contained, its discharge was inaudible at a short distance from the firearm.

Though the cartridges worked well, the pellets' velocity was limited and the round has since been discontinued. Nevertheless, such a round would have definite combat possibilities, especially in covert operations. Unfortunately, the shells are too complicated to be fabricated by reloaders.

MIXED-SHOT LOADS

The mixed-shot load tries to fill in the holes in the pattern that buckshot-only loads create at longer ranges. The idea is that at least a small wound can be achieved rather than a complete miss. The drawback to this is that smaller shot loses its velocity faster than the larger so that, at ranges where the buckshot pattern starts to show holes, the lighter small shot also is none too effective. Nevertheless, the mixed-shot load has met with some success, especially in jungle areas where combat ranges may vary from arm's length to 30 or 40 yards.

The shot should be carefully weighed so an excessive load isn't created; maximum loads should be avoided, since it is nearly impossible to predict what type of pressures will be created with mixed-shot loads.

LINKED SHOT

Linked shot is another attempt to fill in the holes between pellets and keep the shot pattern small at extreme ranges. With this system, a steel wire is used to connect the shot and then the whole thing is loaded into a cartridge. Upon firing, the shot spreads until the wire is fully extended.

Problems with this round include the space taken up by the wire and its increased wind resistance, which quickly slows down the shot. While linked shot seems reasonable in theory, it doesn't work well in reality. If you should try this, the likelihood of creating excessive pressures is very high.

SLUGS AND SABOTS

Slugs fly with the same motion as a dart or badminton birdie. They are not stabilized by the "rifling" on their sides but rather by the fact that most of their weight is toward the front. Because of this, longer slugs will often show some key-holing since they are less ballistically stable in flight.

The sabot round consists of a metal slug smaller than the bore of the shotgun. A pair of plastic "shoes" (sabot means shoe) goes around the slug and seals off the gas so that the whole unit leaves the muzzle when the round is fired. When the saboted slug leaves the barrel, air resistance pushes the sabot halves away from the slug.

Several advantages can be realized by a sabot round. The velocity can be increased and the smaller slug is better able to maintain its velocity because it can have a thinner profile. Upon impact, this thinner projectile can place more energy in one spot for greater penetration and also tends to tumble, creating a larger wound.

While most sabot rounds produced for the shotgun thus far have not taken advantage of its potential increased velocity, it's probably only a matter of time before someone introduces a lightweight, high-velocity, fin-stabilized slug. Fired from a 12 gauge, such a slug could be about the size of a .30-caliber rifle bullet and would, in theory at least, be capable of traveling at 2000 to 3000 feet per second with rifle accuracy. Such a round would certainly give the shotgun the long-range accuracy and power it needs.

Until that time, the combat shooter has to be satisfied with commercial offerings in slugs. A number of manufacturers offer slug ammunition;

all are good. Many shooters feel that the more expensive Brenneke is slightly more accurate.

A large round ball designed for a muzzle-loading rifle will work in a shotgun provided it is encased in a plastic wad column. The plastic wad is needed to seal the burning powder since the ball must be small enough to slip through the shotgun's choke. Lyman makes a .672 ball mold that works well in the 12 gauge, and uses standard lead-bullet casting methods.

Lyman also sells a 12-gauge slug mold with an insert to give the projectile a hollow base. The ball weighs around one ounce depending on the lead alloy you use. Most reloading manuals will show good component combinations for shotgun slugs.

I've seen two oddities in the way of slug modifications. One is to place a metal screw or rod into the top of a slug. In theory this gives it a slightly greater penetrating ability. In practice, it is an easy way to increase the load's pressure dangerously if you don't remove some lead from the slug. It can also cause the shell to set off others in a tubular magazine if the screw sticks up enough to strike the primer of the shell ahead of it.

The other oddity is the "hillbilly" slug. This consists of a standard shot load whose plastic case is cut partway around the side outside the area where the wad separates the powder from the shot. The crimp is then glued shut and/or the shot filled with wax. The idea is that the pressure of firing will rupture the plastic case, and the plastic end of the cartridge—shot, wad, case and all—will go out to hit the target with the lethality of a slug.

It's an intriguing idea but poor in practice. The pressures created if the crimp holds and the shell hasn't been cut quite enough can become dangerously high. If the shell's cut too much, you'll get a blooper. What are the chances of getting it just right? I personally don't care to find out.

Even if you do get the cut right, the shell can't be cycled through a magazine without the whole thing probably coming apart. If you manage to fire it, the hillbilly slug is apt to coat the inside of the choke with plastic. This can become dangerous over time if the barrel can't be cleaned, and you seldom get time-outs for cleaning weapons in the middle of a battle. If you need to fire a slug, get a real slug load.

FLECHETTES

Flechettes are tiny, arrow-shaped projectiles with better ballistics than round shotgun pellets. If someone could come up with a good flechette load for the shotgun, it would increase its range and improve its lethality at all ranges. Flechettes can increase the shotgun's range by reducing the air drag. Since each projectile is fin-stabilized, the flechette round's long-range pattern would be smaller.

Occasionally someone loads short lengths of wire into a cartridge and calls them flechettes. These aren't true flechettes because they tumble in flight rather than traveling point first, and such loads are less effective than round shot since they lose energy quickly to air resistance. Penetrating ability is highly variable at all but close ranges since wire hitting on end penetrates some distance while wire hitting on its side is quickly stopped. Such loads are suitable only for indoor combat and offer little advantage over conventional shot.

SPECIAL PROJECTILES

Tear-gas rounds, grenades, and flares are all available for the shotgun, though generally only in 12 gauge. They work but not all that well, since

12-gauge CS tear-gas cartridge

the shotgun's bore size limits the amount of material that can be delivered. Another option is to mount the tear gas or explosive grenade on the shotgun's barrel and use a special blank to launch it, but generally a tear-gas gun or grenade launcher gives much more satisfactory results.

If a tear-gas projectile is used in a shotgun, it should be remembered that the round is deadly at close ranges, where it has a high velocity. Tear gas should always be launched from a distance toward houses or barricades. The AAI 12-gauge CS tear-gas cartridge is one of the more common rounds. It has fair accuracy within 100 feet but contains only three cubic centimeters of tear gas—hardly enough to be effective except in an enclosed area. Signal flares for the shotgun are usually limited to a burn time of five to seven seconds and an altitude of 200 feet.

RIOT LOADS (RUBBER BULLETS)

This type of ammunition is seen from time to time. The shell's payload contains plastic or rubber balls rather than shot. Because of the reduced power of these loads, they usually won't cycle a semiauto shotgun.

At close range, these rounds can inflict severe injury, but when fired from a distance, usually by bouncing the pellets off the ground so the ricochet hits an opponent, the ammunition can be effective in dispersing a crowd.

The major problem with these loads is that if they are used to control a large crowd, those at the rear may think the police are using standard ammunition. The result may be a panicked crowd smashing through an area to escape or an angry mob attacking officers. Either result could have tragic consequences. Thought and care must be exercised before using these rounds.

SHELL ADAPTERS

Harry Owen manufactures shell adapters which allow a user to shoot .22s, .410 shells, and some pistol cartridges in the 12-gauge shotgun. Though these have little use in combat, they can be used for practice. A .22 shell adapter, loaded with a CCI .22 Long Mini-Cap, would allow very quiet practice even indoors. As a training device, these adapters are hard to beat. The 12 gauge to .410 adapter costs $6, while the pistol adapters cost $25 each. A 6-inch tube for firing .410 shells in 10, 12, 16, or 20 gauge is also available from Harry Owen for $37.

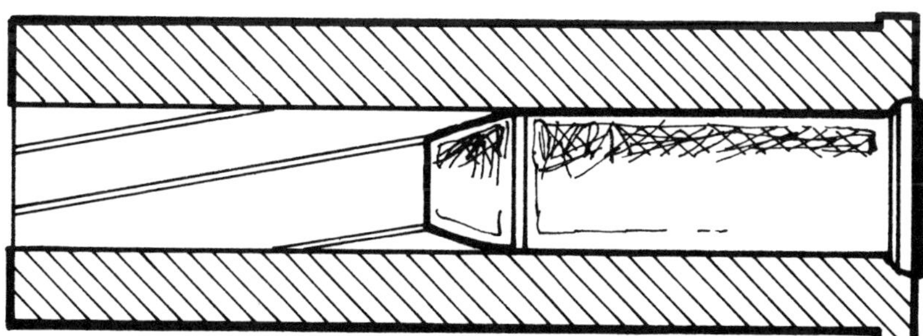

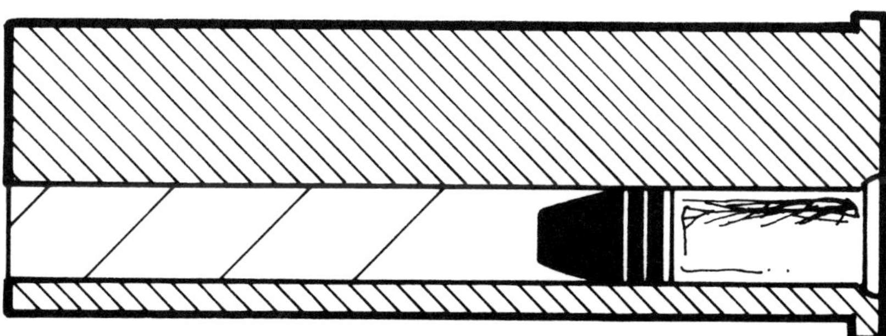

Two Harry Owen 12-gauge adapters are shown. Top: .38 Special to 12 gauge. Bottom: 22LR to 12 gauge

While great strides have been made in pistols, submachine guns, and rifle designs, shotgun design has remained nearly where it was at the beginning of the twentieth century. While the other weapons now have highly effective ammunition, the shotgun staggers along with heavy cartridges that look backward to the age of black powder. As things stand, the shotgunner is at a terrible disadvantage.

Currently, the U.S. military is working on a CAWS (Close Assault Weapon System) which may be a truly modern shotgun. This weapon, if developed as planned, will have a much longer combat range than current shotguns and will fire heavier loads. Whether the CAWS will become operational remains to be seen, but if it does, it's possible that new weapons and ammunition will become generally available. A renaissance in shotgun combat may occur during the early twenty-first century. In the meantime, you should give considerable thought before choosing a shotgun for combat unless it is suited to some unique situation in which you find yourself.

9. When the Stores Are Closed

During both World War II and the Cuban missile crisis, ammunition and reloading supplies were very hard to obtain. It isn't hard to imagine a similar crisis developing again. Likewise, under an oppressive government, laws might be passed to either outlaw ammunition or make it difficult to obtain. One way around this potential problem is to squirrel away a lot of primers, brass, and powder. But even though brass lasts almost forever if you take care of it and bullets can be made from scrap materials, what about powder and primers?

We've already looked at a number of ways to make bullets nearly from scratch. Primers can be reloaded since you can make the primer material as well as smokeless powder. Reloading primers won't become your favorite pastime, but you can create everything you need.

RELOADING PRIMERS

When it looks like reloading components and/or ammunition may become scarce, your first step should be to stock up on both. If you have to choose between buying small primers or large ones, stock up on the smaller ones. They're the hardest to reload. Your next step is to start saving spent primers. Primers are hard to reload, so it is a good idea to have a lot of spare empties in case you damage a few during your primer-reloading endeavors. Commercial primers are easier to recycle than military ones.

One important point: primer chemicals are explosive. Use extreme care when reloading primers and do not keep large amounts of the chemical in the reloading area. Wear eye protection when carrying out this step or any other reloading procedure.

Here's the basic operation:

1. Clean the primers of carbon deposits and dirt.
2. Remove the anvil (the three-pronged, flower-shaped metal insert) from the primer cup. This is most easily done with a sharp-pointed tool.
3. Clean out the primer cup and then use a small tool to flatten out the dent in its bottom made by the firing pin. A hammer with a flat-ground nail works well for this.
4. Place a primer chemical in the cup, compact it into the base, and allow it to dry.
5. Place the anvil in the primer pocket of the brass that is to be reloaded (the brass should be cleaned and polished before the anvil is placed in it).
6. Reseat the primer cup into the brass.
7. Place a tiny paper disc over the primer. The size and thickness of the material will call for a little experimentation. The disc keeps powder from clogging up the primer and preventing the primer "flash" from getting into the cartridge. It may not be necessary if the powder grains are large.

Shotgun primers can also be reloaded. They contain a built-in pocket on the primer itself rather than in the hull, so you will have to dismantle the shot shell primer after it is removed. (A .223 Remington deprimer tool is about the right size for this task.) The anvil in the shotgun primer is a "Y" rather than a three-sided flower; otherwise, the procedure is pretty much the same as with the brass cartridge. It is also essential that the paper disc be placed on the shotgun primer.

Generally, it isn't worthwhile to reload .22 rimfire brass unless it means the difference between having a working firearm or a piece of scrap metal. In rimfire cartridges, the primer goes inside the rim; when the two sides of the rim are mashed together by the firing pin, the primer ignites and sets off the powder.

While you could iron out the firing-pin dents from the inside of the rim, it isn't necessary to do so since the chances of the shell being chambered with the dents exactly lined up are very small. Even if it happens, the shell will still probably fire. Here are the steps for repriming a rimfire cartridge:

1. Clean the old primer residue out of the inside of the rim with a small tool.

2. Dampen the primer chemical with alcohol, acetone, or methyl ethyl ketone (MEK); you'll have to experiment to see what works best. You want it to be almost liquid but not runny so that when you place it in the rim, it will stay there.

3. When you have the right consistency, place the primer material in the cartridge and use a tool to push it into the rim area of the brass. Another way of getting it into the rim is to place the cartridge, base down, into a power drill and spin it at 1000 rpm or so for 30 seconds.

4. Place the brass in a warm place to allow the primer to dry.

Be careful when seating the bullet; the rimfire primer is easily set off during this reloading operation. And remember to wear safety glasses just in case! The bullet in a .22 is held in place by a crimp around the upper rim of the brass. Rimfire ammunition is more sensitive to moisture than other types, so a seal of shellac or lacquer around the rim is essential.

Black powder firearm primers can be made from heavy aluminum foil. The weight found in frozen dinner packages generally works well. The procedure is as follows:

1. Use the nipple of the black powder firearm to form the primer, then cut off the excess aluminum.

2. Coat the inside of the primer cup that you've formed with the primer material.

3. Since many of the primer chemicals are hydroscopic, it's a good idea to keep the black powder primers you make in a sealed container until you're ready to use them.

MATCH-TIP PRIMERS

The tips of strike-anywhere matches are the cheapest primer chemical available. However, most matches currently manufactured will not work because the tip is not explosive enough. This is good news for match users since the match is safer to carry, but it is bad news if you were hoping to use match tips for a primer chemical.

1. Carefully break the white tips off the matches and grind them on a hard, flat surface. A piece of glass works well.

2. Once you have powdered the amount you need, mix it with a little alcohol to create a paste. If alcohol doesn't dissolve the mix, try acetone or methyl ethyl ketone.

3. Place the paste in the primer cup.

4. Let the mix dry fully.

There are a number of other chemicals which can be used for primers: silver permanganate, ammonium iodide, mercury fulminate, antimony sulfide, barium nitrate, potassium chlorate, red phosphorus, calcium silicide, or tetranitroaniline. A number of ingredients are also added. Though some of these chemicals can be mixed together, do not experiment in doing so; you can produce some dangerous results (for example: red phosphorus and potassium chlorate detonate upon contact with each other). Leave mixing operations to the big manufacturers. You can add aluminum flakes or glass powder to the primer mix. Aluminum increases the burning heat, while glass increases the friction created when the primer is struck by the firing pin.

Probably the best primer chemical is barium nitrate since it is noncorrosive (unfortunately, it is often hard to obtain); tetranitoaniline is the worst since it has a short storage life and is very hydroscopic (it attracts water from the air); red phosphorus is also best avoided since it burns on exposure to air and must be mixed while emersed in a liquid. Potassium chlorate is one of the easiest to make and a good place to start.

POTASSIUM CHLORATE

Before making potassium chlorate, bear in mind that it is an explosive. Don't daydream while you're working with it or you may be looking for a new face and fingers. Don't mix it with sulfur or sulfide compounds (it will detonate in a short time) or phosphorus compounds (basically the same results–only more quickly).

To make potassium chlorate, you need potassium hydroxide (lye), hypochlorite solution (household bleach), and sodium bisulfate (in most toilet bowl cleaners). The process of changing potassium hydroxide into potassium chlorate involves substituting the chlorine group for the hydroxide in the potassium compound. The equipment you need is a Pyrex bottle with a one-hole stopper, rubber and/or glass tubing to connect the gas coming from the bottle to a bath in a large pan, a fine cloth or filter paper, and pH paper, litmus paper, or vinegar. Here's the procedure:

1. Make a bath of lye solution in a glass pan (do not use aluminum–the lye will react with it); use one part by weight of lye to two parts of water.
2. Mix several spoonfuls of sodium bisulfate into the bleach; these both go into the bottle.
3. Cork the bottle and vent the gas through the tubing so that it bubbles through the lye solution. The gas produced is poisonous; be sure you vent it fully!
4. When the lye solution is changed to potassium chlorate solution, it will change from a base into an acid. You can tell when the process is finished by testing the solution with the pH paper or litmus paper. If these aren't available, you can test the reaction of a few drops of the solution in vinegar. Basic chemicals fizzle when placed in acids like vinegar. When the fizz is gone, the solution is acidic.
5. When the solution has become acidic, place it into a shallow pan and place a small flame under it. *Do not* allow the solution to boil violently.
6. Watch for crystals forming on the bottom of the pan. When this happens, turn off the heat and allow the solution to cool. The potassium chlorate will precipitate out of the solution.
7. When the crystals have precipitated out, pour the liquid and crystals through a cloth, paper towel, or other filter. Save the crystals which are trapped in the filter.
8. Dissolve the crystals in water, one part crystals to one part water.
9. Place the mix into a shallow pan and place a small flame under it. *Do not* allow the solution to boil violently. The crystals should vanish as they dissolve. (The next few steps purify the potassium chlorate.)
10. Continue heating until crystals again form on the bottom of the pan. When this happens, turn off the heat and allow the solution to cool so that the potassium chlorate again precipitates out of the solution.
11. When the crystals have precipitated out, pour the liquid and crystals through a cloth, paper towel, or other filter. Keep the crystals trapped in the filter.
12. Grind the potassium chlorate crystals to the size you need while they are wet. Follow the directions for working with match tips. Make only small amounts since this chemical is highly explosive.

MERCURY FULMINATE

Making mercury fulminate is also a simple procedure but calls for mercury and nitric acid. These chemicals are usually hard to obtain, although, as we'll see later in the powder section, it is possible to make the nitric acid. Follow these steps:

1. Mix 10 parts of mercury to 74 parts by measured volume of nitric acid.
2. When the mercury has dissolved, heat the solution to 130°F.
3. Slowly pour the solution into 100 parts by measure of alcohol. Be careful in pouring since splashing can create dangerous burns or fire. An effervescent reaction will occur, and white fumes will be given off. The fumes should be vented since they are dangerous.
4. When the reaction stops, filter the liquid through paper towels or filter paper.
5. Wash the particles trapped in the filter with cold water.
6. Place the particles in a pan and dry them by placing the pan over boiling water.

AMMONIUM IODIDE

Iodine crystals can be purchased at drug stores and ammonia at supermarkets. The main drawback with making this chemical is that it is time-consuming.

1. Break iodine crystals into approximately 1/8-inch-diameter pieces.

2. Place two parts by volume of iodine crystals into a glass container and place three parts by volume of ammonia over them.

3. Allow the mixture to sit for three days.

4. Pour off and discard the liquid, carefully removing the crystals in a cloth or paper filter.

5. Again place two parts of the crystals removed by filtering into a glass container and place three parts ammonia over them.

6. Allow the mixture to sit for three more days.

7. Pour off and discard the liquid, carefully removing the crystals in a cloth or paper filter.

8. Place two parts crystals left into the container and place three parts ammonia over them.

9. Allow the mixture to sit for three more days.

10. Pour off and discard the liquid, carefully removing the crystals in a cloth or paper filter.

11. Allow the crystals to air-dry in a well-ventilated area. Do not expose them to heat.

12. The ammonium iodide is now ready to be ground into powder. Like all other primer chemicals, this compound is very explosive and should be treated accordingly.

When you've mounted the primer in the brass, it is important to seal it in place with shellac or lacquer around the outside rim. Most of the do-it-yourself primer chemicals are hydroscopic and will pull water out of the air if exposed to it.

After firing a weapon loaded with any of these primers, clean it immediately; the primer chemicals are corrosive and/or hydroscopic. Failure to do so will cause rust inside the barrel. Probably the best method of cleaning out a barrel is to pour boiling water down it several times and then follow up with standard bore-cleaning procedures.

POWDER

A number of powders can be easily manufactured. All are dangerous to varying degrees; the greatest room for disaster lies, however, in making modern powders. One mistake and you can create a serious explosion, fire, or acid spill. Making powder is extremely dangerous.

Some of the more common propellants are black powder, ammonpulver, cordite, guncotton, and potassium nitrate mixtures, or combinations of these. Storage can be a problem since most powders are hydroscopic to some extent, may decompose over time, and some, especially black powder, are more akin to explosives than propellants. Whenever possible, keep your powder in a cool place in a tightly sealed container. The container will rupture if exposed to a fire and could become a bomb!

POTASSIUM NITRATE POWDERS

The simplest powders to make are created from potassium nitrate (saltpeter). The most common is created by mixing seven parts by measure of potassium nitrate to six parts of sugar. Potassium nitrate powder works well with rifles under .50 caliber. It doesn't work well in shorter barrels like those of most pistols or submachine guns but does work in long-barreled carbines using pistol cartridges.

Animal droppings (bats and birds) which aren't disturbed for long periods of time are broken down by bacteria. One of the chemicals created in this process is potassium nitrate. The catch is that this process works best at around 100°F and large amounts of time and dung are needed. Thus, natural sources of potassium nitrate are usually restricted to caves in warm climates.

A more roundabout route to speed up the process involves chemically binding calcium with nitrate and then exchanging potassium for the calcium.

To carry out this process, place lime in an outhouse pit which has been in use for some time. Dig up the pit and mix the earth/excrement/lime in water so that the calcium nitrate dissolves out of the earth mix. Remove the water from the material that settles to the bottom of the mixture; in so doing, you have dissolved the calcium nitrate. Boiling the water leaves behind the calcium nitrate and, unfortunately, some other salts as well.

The next step is to add wood ash, which contains potassium carbonate, to water and then dissolve the calcium nitrate crystals in the solution. The potassium and calcium exchange places, creating calcium carbonate (quicklime) and potassium nitrate. Filter the solution to remove the calcium carbonate, which is more or less solid; the potassium nitrate dissolves in the water. Filter paper can be used, but several passes through a coarser material work better since the calcium carbonate quickly clogs up most filters. Strain filters were used in the 1800s; this still works well. The water should be filtered a number of times to get as pure a solution as possible.

To retrieve the potassium nitrate, boil the water

until it is nearly gone. Before the water totally evaporates, remove it from the heat and allow it to complete evaporation by leaving it exposed to the air. The crystals left behind will be mostly potassium nitrate.

Below is a more modern step-by-step procedure that yields better results:

1. Get a bucket and punch holes in its bottom.
2. Place a cloth over the bottom of the bucket; place straw over it and add another layer of cloth. Next, add another layer of straw followed by a third layer of cloth.
3. Place a large container under the bucket.
4. Fill the bucket with your source of nitrate and earth.
5. Pour boiling water through the soil and let it drip into the container beneath the bucket. Pour the water very slowly and use only one part water for every two parts earth mix.
6. Allow the water to stand for a number of hours and then drain it off and discard any sediment which may have gotten through the filter.
7. Boil the water for two hours. While the water is boiling, small crystals of salts will form on the bottom of the container; remove these and discard them.
8. When half the water has boiled off from the solution, let it cool for half an hour.
9. Add one part alcohol to one part solution and filter the mixture through a paper towel. Crystals of potassium nitrate will be left in the filter.
10. To purify the potassium nitrate, dissolve it in clean water and boil the water for two hours. While the water is boiling, small crystals of salts will form on the bottom of the container; remove these and discard them.
11. Again add one part alcohol to one part solution and filter the mixture through a paper towel. Crystals of potassium nitrate will be left in the filter.
12. Allow the crystals to air-dry, then seal them in a waterproof container until they're needed.

Once ample supplies of potassium nitrate have been created, it is necessary to mix the potassium nitrate with sugar to create the actual powder. Here's how:

1. Mix 56 parts by volume of potassium nitrate with 48 parts sugar in 84 parts of water. (Up to three parts ferric oxide [rust] are sometimes added.)
2. Boil the mix over a small flame while stirring it. Both the potassium nitrate and sugar should dissolve in the water.
3. Boil the water down to one fourth of its original volume. This should create a thick "fudge."
4. Pour the mixture out on a flat surface and allow it to dry in the sun. Scoring deep lines or furrows in it will speed up the drying process.
5. When the mixture is moist to the touch but no longer sticky, granulate it by pushing it, a little at a time, through mesh or screening.
6. Allow the particles to dry in the sun.

Your powder is now ready to use. Store it in an airtight container or reload it, and carefully seal the cartridges since the mixture is highly hydroscopic.

POTASSIUM CHLORATE

Potassium chlorate can be used as a black powder substitute as well as a primer material. As a powder, it must be mixed with sugar to slow down its burning rate; the mixture can then be substituted for gunpowder. Use slightly less potassium chlorate powder for the same load than black powder. In a bind, this material can also be used in smokeless cartridges but may not create enough energy to cycle semiauto weapons.

Safety matches are almost pure potassium chlorate; on strike-anywhere matches, be sure to remove the tips. Failure to do so could create excessive chamber pressures.

You'll need a candy thermometer or the like to keep track of the temperature of this mix. Note the following steps:

1. Mix one part table sugar in just enough water to dissolve it into a slush and heat it slowly to 250°F until the sugar melts. Take care not to let the sugar turn brown from excessive heat.
2. Remove the sugar from the fire and stir it as the temperature drops.
3. When the temperature of the sugar has dropped to 150°F, add one part potassium chlorate to the fudge a little at a time.
4. Pour the mixture out on a flat surface and allow it to dry in the sun.
5. When the mix is dry, granulate it by pushing it through mesh or screening. If the mixture is hard, too much heat has been applied to the sugar; if the mixture is gooey, too little heat was used. In either case, discard the mix and start over, paying careful attention to the temperature. The size

of the granules determines the desired burning rate. Small granules should be used for pistols or shotguns, while larger sizes are suitable for muzzle-loading rifles.

Your powder is now ready to use. Seal it up or reload it and carefully seal the cartridges, since the mixture is highly hydroscopic. Don't forget to clean your firearm after using this powder.

BLACK POWDER

Black powder is both hard to make and dangerous. Sulfur is hard to come by in most areas, and the material is explosive when dry. Because you need potassium nitrate to make it, it is easier to just mix the potassium nitrate with sugar to create the powder described above, which is basically black powder with the fuss and sulfur left out. But if you must make black powder, here is how:

1. Obtain potassium nitrate (see above) and carefully grind it into powder.
2. Get some charcoal and grind it to powder.
3. Grind sulfur into powder.
4. Mix the three powdered chemicals in the following amounts by weight: 50 to 75 parts potassium nitrate, 8 to 30 parts charcoal, and 10 to 15 parts sulfur. Generally the more sulfur, the faster the burning rate; increasing the sulfur to 20 parts will make it unsuitable for firearms but useful as an explosive.
5. Add water to the powder and mix it together thoroughly. As long as the mixture is wet, it will not explode.
6. Allow the mixture to dry and then carefully break it into chunks. (Be extra careful; it is explosive at this point.)
7. Grind the chunks up one at a time. Protect the ground powder and unground chunks from accidental ignition. A screen or sieve is ideal for the grinding process. Fine powder is used for smaller firearms, while coarser grains are used for rifles.

The powder is now ready for use. It should be stored in airtight containers away from your living area since the powder is explosive. After firing a weapon charged with black powder, clean it thoroughly to keep it from rusting.

AMMONPULVER

Ammonpulver is made from ammonium nitrate. This chemical is commonly found in agriculteral fertilizers and readily available in most areas. Wood alcohol is needed to process the ammonium nitrate, and charcoal is mixed with the chemical to create the proper burning rate.

1. Place the fertilizer in a pan and pour methanol (wood alcohol) over it until the fertilizer is completely covered. The mix should be kept warm, but don't apply heat to it.
2. Stir the mix until most of the fertilizer is dissolved in the alcohol. With coated fertilizers, this may take some time.
3. Pour off the alcohol and save it; discard the sludge left behind.
4. Cool the alcohol by placing its container into an ice bath or, better yet, into a container of dry ice. This will cause crystals of pure ammonium nitrate to form in the container.
5. Run the alcohol and crystals through a filter. The alcohol can be reused for retrieving ammonium nitrate from another load of fertilizer.
6. Place the crystals in the sun and allow them to dry.
7. If the crystals are not small enough, grind them with extreme caution.
8. Mix eight to nine parts crystals with one to two parts charcoal.

This mixture is highly hydroscopic and should be kept in sealed containers. Ammunition made with it should also be sealed. Always clean firearms after firing this ammunition to prevent excessive corrosion of the barrel and other parts.

NITROCELLULOSE (GUNCOTTON)

The procedure for making nitrocellulose is dangerous. Sulfuric and nitric acids are needed, both of which are highly caustic and can be contained only in glass without creating dangerous reactions. They are hard to neutralize; a spill will call for a lot of water and a strong base (like soda or lye) to clean things up and make them safe again. Gloves, safety goggles, lab apron, hooded ventilation systems, etc., are all absolutely necessary for this work.

In addition to the two acids, you'll also need a relatively pure source of cellulose, the best source of which is probably cotton. Wood pulp or tissue paper might also work but usually are not satisfactory.

Since you may not be able to find nitric acid, we'll take a look at how to create it with sulfuric acid. If you have both chemicals, skip the first five steps. Diluted sulfuric acid is commonly used in vehicle batteries. To concentrate it, boil it until white fumes appear. (These fumes are dangerous; adequate ventilation is essential.) When the fumes start, lower the heat since the acid is ready for use.

1. Prepare dry potassium nitrate using the above-described method.
2. Place two parts of the potassium nitrate crystals into a lab bottle with a one-hole stopper. Place a piece of glass tubing with rubber tubing attached to it in the hole. Don't fill the bottle more than one fourth full of the crystals.
3. Add one part sulfuric acid to the bottle and mix with a glass rod until the crystals form a paste.
4. Run the rubber tubing through an ice-water bath and into a second open bottle. Place the stopper in the bottle with the potassium nitrate/sulfuric acid paste in it.
5. Place the paste bottle over a gentle heat. This will create nitric acid fumes which will condense into a liquid when they go thorugh the bath. The nitric acid will then drip into the second bottle. Take extreme care not to breathe the fumes that do not condense in the bath, and be sure that you have adequate ventilation.
6. Immerse the cotton in a mixture of one part sulfuric acid and one part nitric acid. Allow the cotton to become saturated, then remove it and store it in a sealed container for 24 hours in a cool area.
7. Carefully rinse the cotton in pure water until it gives a neutral reading with litmus paper or pH paper. If it is the least bit acidic, rinse it again.
8. Dry the cotton, which is now nitrocellulose. Take extreme care to keep the temperature of the nitrocellulose below 212°F. Once dry, the nitrocellulose must be stored at a cool temperature and not exposed to sunlight.

At this point, you may process the nitrocellulose for use as a powder by dissolving it into a gelatinous blob in ether or alcohol, then rolling it into sheets to be cut or molded for use in ammunition and then redried.

MODERN SMOKELESS POWDERS

Smokeless powder is created from nitrocellulose and nitroglycerine. In making nitroglycerine, you'll need glyercine, nitrocellulose, sulfuric acid, and nitric acid. Again, remember that these procedures are extremely dangerous since sulfuric and nitric acids are used. If you have not read the precautions outlined in the section on nitrocellulose, go back and read them *now.*

Warning: In following the steps below, you will create nitroglycerine. Make only small amounts, and do not jar the chemical or allow the chemicals being used to make it to rise above 86°F. If the mixture rises above 86 degrees, an extremely large explosion will result.

After completing steps 1 through 8 in the preceding section, continue as follows:

1. Mix 100 parts of nitric acid with 200 parts of sulfuric acid in a water bath which will keep the containers below 50°F. These acids create heat when agitated, so take extreme care to maintain the mixture's temperature.
2. When the temperature of the mixture has stabilized, *very* slowly add 38 parts of glycerine. If possible, allow it to trickle down a glass tube rather than dropping down into the mixture, which could cause a violent reaction.
3. *Slowly* stir the mixture with a glass rod for ten seconds. Stirring can create heat; be sure to maintain the temperature below 50 degrees, or a violent explosion will result.
4. Gently pour 340 parts of water into the mixture.
5. The nitroglycerine will precipitate out of the mixture. Carefully remove the nitroglycerine from the solution, taking extreme care to keep it below 50°F. *Do not* jar or drop it.
6. *Very gently* wash the nitroglycerine several times with water, and use sodium bicarbonate or lime to neutralize any acid in it. (Failure to neutralize the acid will give the powder a very short shelf life.) Keep the temperature of the nitroglycerine very low.
7. Dissolve the nitrocellulose and nitroglycerine in a solvent and extrude or mold it to the correct size to burn properly and create the correct pressure curve for the firearm to be used. A good mix is 58 parts nitroglycerine, 37 parts nitrocellulose, and 20 parts acetone.
8. Wear rubber gloves, gently knead the chemicals until they are well mixed. At this point, five parts petroleum jelly may be added to reduce barrel erosion. If it is available, carbazole, also known as diphenylene amide or diphenylamine, can also be added to the powder mixture to help stabilize it.
9. Extrude the mix through a sieve or screen so that it forms threads. The smaller the sieve size, the faster the powder will burn. Large-size threads are suitable for rifle ammunition and smaller

threads for pistols and shotguns.

10. Allow the threads to dry at very low heat and then break them into very short lengths.

This smokeless powder is relatively safe to handle but will be heat and light sensitive. It should be stored in a sealed container at cool temperatures.

Most of the powders and primers listed above require cleaning the rifle shortly after firing. Failure to do so can cause extensive rusting in the barrel or in the gas system of gas-operated semiauto weapons. Remember, however, that noncorrosive primers and powders are a fairly recent development; many old timers will tell you that properly cleaned firearms last just as long and are every bit as effective with corrosive powders.

BRASS

Brass cartridges are next to impossible for an individual to cast or swage. Though straight-walled cases might be made using a lathe, their walls would be so thick that they would have to be used with squib loads to avoid creating dangerous pressure.

The best route is to carefully save fired brass; with semiautomatic weapons, a good brass catcher makes sense. These make it possible to use a weapon without worrying about lost brass. While they are not normally recommended for combat use, in a situation where you were running out of brass to reload, the use of a brass catcher would probably be warranted. (The best catchers are currently being made by E & L Manufacturing. They have catchers for the Mini-14, AR-15/M16, H&K 91/93/94, Uzi, MAC 10/11/11-9, .30 Carbine, KG-9/99, Remington 1100, S&W 1000, and Remington Four/74 among others. The catchers cost $25 each, and the shotgun catchers hold 5 rounds; others hold 60 or more rounds.)

To reload the brass, use the methods outlined elsewhere to maximize case life and, whenever possible, use lighter loads. Trimming, annealing, and careful checking of the cartridges are essential to prevent damage to your firearm and yourself.

When brass has been fired and reshaped a number of times, it becomes somewhat brittle and may crack or split. It is possible to anneal the brass neck to extend the case life. Annealing must be done carefully and the head of the case cannot be softened or it will become dangerously weak.

To anneal the neck, clean the cases so you can gauge the temperature by color change, darken the light in the workplace somewhat, and place the cartridges in a pan of water up to each case's shoulder. Heat each cartridge neck with a blow torch or similar device until the brass starts to darken. The case should not turn cherry red; continue to heat the neck for three seconds or so after the case becomes dark. When this time has elapsed, knock the cartridge over so that the mouth is cooled by the water; this gives it the proper hardness for reloading. Take your time heating each case, and try to spend the same amount of time on each one so they'll all have similar softness. The water will keep the case head from becoming too soft. Be sure the cases are dry before reloading.

If you have a large number of cases to anneal and have a lead melting pot with a temperature control, the following method may be used for annealing. Set the temperature control between 800 and 850°F, and allow the lead in the pot to melt. After giving the brass case a light coat of oil, hold the head in your bare fingers and dip the cartridge into the lead up to the shoulder area. When the brass starts to feel warm, toss it into a container of water. Once your cases have been annealed and cooled, remove the oil with detergent and set them aside to dry. If you dry them in an oven, do not allow the temperature to rise above 175°F. Very few ovens are accurate enough to allow for this precise temperature. Temperatures above 200 degrees will ruin the cartridge head's strength.

Generally, you'll be able to anneal cartridges twice before running into problems. These will give you a large number of reloads if you are careful to use light loads.

When brass has been used a maximum number of times, it can usually be used for squib loads. Through such recycling, the useful life of the brass can be extended. Care should be taken when using mercuric primers; they cause brass to become brittle with time after being fired.

It is also possible to alter ammunition to work in other calibers. For example, the .30-06 or .308 Winchester cartridge can be cut down to accept .45 ACP bullets, and the .223 Remington and similar cartridges can be cut down and expanded into 9mm Luger, .380 ACP, and maybe .32 ACP or .30 Carbine cartridges. If a little careful hammering is done on the base to create a rim, the .223 cartridge can also be used in .38 Special or .357 Magnum chambers, and the .308 in the .45 Colt, and with great luck, the .44 Magnum. This is a very crude adaptation, and the thicker brass of the cases requires that they not be loaded anywhere near the maximum used in standard cases; excessive pressures and possible cartridge rupture would be sure to result. It's also a good idea to use such brass with a squibbish load at the first firing so that the case can be fire-formed to size. In order to make the altered case's rim fit the firearm's extractor, it may be necessary to place the brass in a lathe or drill and use a file to change its shape (be careful not to weaken the case too much). To cut the brass to size, a plumber's tubing cutter will work well; a file and brass trimmer will finish the job. When adapting brass to a slightly larger size, such as the .223 to 9mm Luger, it may also be necessary to make an expanding rod.

No pistol brass created by the above method will be very reliable, especially in semiauto actions, and won't be 100 percent safe. Use such methods only in an emergency.

If you don't have resizing dies, you can resize empty brass by forcing it into a firearm chamber. While a firearm makes a rather mediocre reloading tool, it may work in a pinch. Oiling the brass during the sizing operation will save wear and tear; just remember to wipe it off the brass and out of the chamber before firing the round to prevent excessive backward pressure or gas leakage.

Warning: Some really dangerous practices which will sometimes result in a ruptured case and escaped gas are described below. These should be tried *only* if you will die if you do not fire your weapon.

One such practice is to load the .223 Remington into a .222 Remington Magnum rifle. Another is to load the .308 Winchester cartridge into a .30-06 chamber, which will sometimes work if the .308 is larger than the chamber. The brass will probably be ruined in any case and either round is quite dangerous in the incorrect chambering. But it might save your life if you need to fire the rifle in a combat situation; weigh the risks before pulling the trigger.

With shotguns, things are more limited. Either .30-06 or .308 brass can be cut to the proper length and a rim peened on it for use in a .410 shotgun. A 16-gauge shell can be made to fit in the 12 gauge by wrapping the smaller shell with tape or paper; this is not overly safe, however, and 16-gauge shells are usually less common than 12 gauge.

With pistol, rifle, or shotgun ammunition, probably the best measure is to have lots of spare cases and components. Shell adapters like those from Harry Owen are a good idea, too. In addition to allowing you to fire lighter loads, they allow you to fire scrounged ammunition in a pinch. They do this by adapting a variety of smaller shells with the same size bullets for use in firearms. These adapters cost only $17 each and will be worth their weight in gold—or ammunition—if the stores close for good.

BULLETS

With a few good bullet molds and/or some swaging equipment, you could probably keep yourself in bullets with only a little scrounging around for materials. As noted in other sections of this book, it is also possible to create bullets by turning brass or copper rods. Plastic and wood can also be employed for short-range bullets; either type is lethal within 20 yards or so.

Generally, it is better not to try to place a coat on the lead bullets in an effort to increase the velocity at which they can be used without leading the bore. While the idea sounds good, it is next to impossible to implement. Tin or other metal washers should never be used since they can chemically weld themselves to the brass cartridge over time and create excessive pressures when fired.

Two possibilities might be of use. One would be a copper coat added by electrolysis. This is very iffy, however. A better way is to use Corbin's .22 rimfire jacket maker to transform .22 LR empties into .224 bullets. Though this method is tedious, it does work. The only drawback is that the bullet jackets aren't quite as strong as copper jackets and should not be used in .223 Remington rifles with faster twists.

In a real bind, brass or steel bolts or metal screws with their heads cut off could be seated in the brass; barrel wear would be excessive, but the risk might be worth it if you are completely out of bullets. If the bolt is a bit too small to fit tightly in the brass (which should be resized—by the firearm chamber if necessary), tape or paper can be wrapped around the projectile before it is seated.

A wide range of bullets can be used with cartridges other than those they were designed for. Weight can often be removed with a little judicious filing, drilling, or knife work to create a lighter bullet for smaller cartridges. Lightweight bullets in heavier loadings usually are no problem except for lead bullets, which may create leading problems if they are propelled too fast.

In general, the .357 Magnum, .38 Special, .380 ACP, .32 ACP, and 9mm Luger (or other 9mm cartridges) can all use the same bullet if the weight is reduced for the smaller cartridges. Though the .357/.38 bullet is .002 of an inch larger in diameter than the 9mm, the barrel will swage the bullet to fit with only a slight increase in chamber pressure; this is a problem only if very hot loads are used, so be sure to avoid them. Be careful rather than sorry.

The .308 Winchester, .30-06, .30-30, .30 Carbine, and .300 Magnums can all use the same diameter bullets; the principle difference is weight. Likewise, .224-diameter bullets work in a wide range of .22 centerfire cartridges like the .223 Remington, .222 Remington, .222 Remington Magnum, .22 Hornet, .22 LR, etc. Except for the .22 LR bullet, which is quite close to the same diameter, all these rounds use a .224-diameter bullet; again, bullet weight and avoiding maximum loads are the main considerations.

In rifle and high-speed pistol cartridges like the .357 Magnum and 9mm Luger, it should also be remembered that lead bullets cannot be used with maximum velocities without severe leading problems. In such a case, it is better to use heavier, blunt, and/or expanding bullets propelled at a slower velocity for maximum effectiveness without ruining accuracy or increasing pressure levels in the barrel because of leading.

IT CAN BE DONE

It would be tough going, but you could manufacture your own ammunition practically from scratch if you had to. Knowing how could save your life by allowing you to use a firearm when regular ammunition supplies were exhausted. In addition, being able to create ammunition during such a period would be a fantastic barter skill, albeit possibly illegal. Provided you have some time and raw materials, you can do it.

10. Combat

Combat takes its toll on your body as well as your mind. Neither mind nor spirit operate the same way in combat that they do in everyday life.

When combat nears, you may become physically ill; diarrhea and vomiting are not unusual among troops awaiting battle. As danger approaches, a sudden rush of adrenaline will enable you to do superhuman feats; blood is routed from less important organs to your brain and muscles so you will have no feeling of hunger or other discomfort until afterwards. The adrenaline in your body makes fine muscle control vanish; your hands will grip your weapon as if it were a baseball bat. Simple tasks like aiming or reloading your firearm will become nearly impossible. Your hands will probably shake uncontrollably. These are not signs of fear; this is how your body gears up for survival.

Your mind pulls some strange tricks, too. Vision often seems to narrow in combat so that you see only what you're shooting at. Peripheral vision may actually vanish. With this tunnel vision you may not be aware of what is happening to either side of you. Subconsciously you'll assume a combat crouch to present less of a target.

If you've trained and practiced carefully, your limbs may function on their own. You may feel as if you're totally out of control and only watching as some other being directs your actions (it isn't rare for a shooter to empty his firearm and not even be aware that he was shooting). Because of this "auto mode," practice, along with a good weapon fed with reliable ammunition, can give you the winning edge in combat.

Realistic practice is essential. You should practice on human-shaped targets, not bull's-eyes. Practice point shooting as well as sight shooting. Try not to switch from one weapon to another a lot when you practice—chances are you'll get them confused in combat. I've spoken to men who have tried to release the Garand safety when they were carrying an M1 Carbine; they wasted precious time getting their shots off and were lucky to survive.

Do some vigorous exercise before you try out combat-shooting techniques so that you'll have a quasi-adrenaline high. Practice with your muscles twitching as you gasp for air—that's how you will feel during actual combat. Develop the habit of looking around as well as straight ahead so that you'll overcome any tunnel vision you may experience during combat.

Don't forget armchair practice. Think about various combat scenarios and how you'd counter them. Think about what you'd do before, during, and after a confrontation with an enemy. What happens when you've downed one foe and another springs out of hiding? Don't limit yourself to a one-on-one fight. Think things out before the fact and you'll be much better prepared to deal with a real battle.

Give yourself an edge with the weapon and the ammunition you use. If you have any idea of where you'll be fighting, try to arrange things so that the environment works for you, not against you. Select your weapon and ammunition to take advantage of the environment or try to use tactics that will allow you to use the edge your ammunition may have over that of your enemy. For example, if you have a .308 and the enemy is armed with a .223, try to use your long-range capability. Shotgunners should try to fight in areas where they have minimal distances to cover. Think and plan.

Know the difference between concealment and cover. Concealment hides you from sight but gives no protection. Cover gives you bulletproof protection from an enemy's fire, depending on what your opponent is shooting. When you examine a potential combat area, figure out what is cover and what is only concealment. If your enemy makes an error, you may be able to nail him when he thinks he's safe. A lot of combat ends with shots that are fired where the winner suspects his enemy is. Don't always wait for a clear shot.

If you are unsure about what your ammunition can do, one excellent video tape used by the U.S. military and the FBI in training is entitled "Deadly Weapons," available in VHS or Beta from the Anite Company. The tape covers the penetration of most of the combat calibers on cars, etc., among other things, and gives hands-on experience in the privacy of your living room.

Finally, you have to be sure of what you're doing when it comes time to actually pull the trigger to defend yourself. If you hesitate or have doubts, chances are good you won't survive. You're better off running or surrendering.

PISTOLS, SUBMACHINE GUNS, AND SHOTGUNS

The moment of truth for you and your combat ammunition comes during what I call a "close encounter of the worst kind." This is the close-range shoot-out with an armed opponent. Generally a pistol, submachine gun, or shotgun will be used by one or both sides, and the fight can occur during "everyday" crimes as well as actual war conditions.

To prepare for such an encounter, forget anything that Hollywood movies may have taught you. Forget hiding behind "bulletproof" tables and overstuffed furniture that provide only concealment, not cover. Cover is very rare in a modern home.

The other thing to forget is the instant kill of TV combat. Often an opponent won't be instantly downed even by a fatal shot or, if he falls, he may not be seriously wounded. Assuming that your job is done when the bullet connects with your target can be a fatal error when your foe shoots back.

While the Hollywood cop or soldier kills an opponent, makes a witty remark, and goes back to business as usual, the real-life hero may suffer psychological problems from close encounters. These will be less of a concern when you're defending home turf or are clearly in the right. But the close-combat encounter can be very damaging psychologically. Don't hesitate to get professional help if you start to have a few doubts or emotional ups and downs. A large percentage of combatants in "everyday" battle need professional counseling the same as do those in military combat. Guys who try to "tough it out" often later become casualties either through careless accidents or psychological upheavals.

An "everyday" shooting situation and its aftermath is worse than a similar occurrence on the battlefield. On the battlefield you are often treated as a hero for winning the encounter. In an "everyday" fight you'll be treated like a criminal by police, community, anti-gun propagandists, and possibly newsmen. You may even discover your "friends" are not such friends after all. This often continues after your innocence has been proven to the satisfaction of the courts. Following a fight, you'll probably be treated like a criminal and will have your firearm confiscated as "evidence." You'll probably spend a night or more in jail. Now is the time to become mentally prepared to deal with these problems so that they don't surprise you if and when they happen.

When fighting in "everyday combat," you need to remember that innocent bystanders may be outside the walls of the area in which you're fighting. Use expanding ammunition and exercise caution about where you fire—a 9mm hollow-point bullet will go through five layers of one-half-inch plywood while a .357 hollow point will go through four and one-half inches. If at all possible, never use a pistol loaded with FMJ ammunition, a shotgun with buckshot, or a rifle indoors when an exiting bullet might injure someone. Because most combat-pistol ammunition will penetrate several walls in most houses, the best tactic is to position yourself so that you'll be firing in a relatively safe direction. Again, such tactics must be planned out well in advance of any actual fight.

Don't forget that a gun battle isn't over until someone has been killed or injured to the point where he can no longer fight. Many criminals now wear ballistic vests and have concealed weapons. Don't take chances; be very sure the fight is over and all your opponents are down for the count or have retreated.

If you take prisoners, remember that they may not be docile. Surrender may only be a ruse to get close enough to shove a knife blade between your ribs or make a grab for your weapon. Keep prisoners facedown on the floor until everything is under control.

Never allow an enemy, armed or unarmed, to approach too near. If he fails to stop after a warning, you should assume he has a concealed weapon or will try to engage you with his fists. Since you have no way of knowing his skill or training (does he have a black belt in karate?), he may be able to win such a battle, given the chance. You must fire on such a person if he refuses to desist or you will risk losing your life and the lives of those who depend on you for protection.

If the police will be on the scene in the aftermath of a battle (or if you are out of uniform in combat), be sure "friendlies" don't mistake you for the enemy. Don't run out to greet the police or troops with a gun in your hand unless you're interested in doing a Swiss cheese imitation.

Close-range combat, where you actually see your enemy, is extremely traumatic. You'll need to plan and train carefully to survive a close encounter of the worst kind.

COMBAT WITH RIFLES

There are two views of what is necessary in combat. One is that a fighter should use a long-range rifle to take careful, precise shots at an enemy. The other view—which is the current trend—is that firepower, a massive barrage of lead, is more effective than carefully aimed fire. There is truth in both viewpoints, but what you do will depend on your weapon training, and combat experience. Failure to be realistic about this will probably be fatal to you.

The ability to hit long-range targets will not necessarily help you in combat. Combat with rifles requires fast shots at fleetingly viewed opponents, and most military studies show that the best way to hit a fleeting target is to fire a number of bullets in its general direction. But if you're using a weapon chambered in .308, .30-06, etc., then you'll have to buck the system and try to use tactics that will allow you to use carefully placed shots rather than a barrage.

With smaller calibers of weapons, the three-round burst is actually more lethal per shot fired than full-auto fire. Many maintain that semiauto fire is the best mode to use if a three-round burst control isn't available. This is because the rifle goes out of control after the first several shots in a long burst, making a quick follow-up impossible. The three-round burst feature or semiauto fire prevent emptying your weapon during the panic of combat.

Most modern battle tactics are based on movement. Movement can be achieved more easily if an enemy is pinned down by suppressive fire. According to modern military principles, suppressive fire can be achieved with 40 to 100 rounds per minute. Most semiauto .223 rifles or carbines can handle this amount of fire for a short period.

Those who use the larger .30-caliber weapons or are limited to semiauto fire may be able to increase the odds in their favor with multiple projectile rounds or new aiming systems such as laser or aimpoint scopes. With the larger .30-caliber weapons, the first shot must be accurate because the follow-up shot takes longer and, once warned, your opponent is apt to be returning a barrage of fire.

Regardless of what caliber you use, you usually have to locate an enemy before you can eliminate him. There are a lot of elements to watch for, including muzzle flash (in low-light conditions), sounds, reflected light from shiny gear, etc. To see or hear these things, you must place yourself in an area where you have maximum visibility with full cover or concealment.

When you search for an enemy, first examine the area nearest you to prevent any unpleasant surprises. Spend about thirty seconds searching until you gradually work your way out to 300 yards, which in most areas is the combat-range limit unless you are atop a hill, overlooking a plain, etc. Care must be taken when looking for an enemy. He'll be camouflaged and easily missed. Focus your search on specific spots which could provide concealment. If an enemy isn't located during the thirty-second search, go over the area visually in a slower search.

When the second search is completed, maintain a watch over the area, studying anything that is suspicious or unusual. Learn to "think like an enemy" to locate potential hiding spots. If you're in an area you're familiar with, give some thought as to where you would hide and then look at those spots.

Your life may depend on how well you are hidden. Avoid quick movements and be careful to avoid creating reflections with metal gear and optical equipment. Be sure your camouflage technique is adequate and *keep quiet*—most combatants make enough noise to be detected from 100 yards when they move.

An enemy probably won't remain exposed for long. Always take note of what is around the area he's in so that you can hold your fire until you ascertain his numbers. The best way to keep track of your enemy is to mentally mark the spot he's in. To do this, use either an aiming point or a reference point.

An aiming point is an object the enemy is in or behind. A reference point is some object near an enemy's position, like a bush. A reference point is a bit harder to use since you must estimate how far behind, in front of, or to the side of the object the enemy is located. Pick easily located aiming or reference points so that you can find them again or tell a comrade where they are.

While most combat rifles with a 250-yard zero are virtually flat-shooting within battle ranges, if you're sniping or using a .30-caliber rifle and wish to take advantage of its range, you'll need to be able to estimate the target distance to compensate for bullet drop. One easy way to do this is to use a range-finding scope. While these demand extra care, you should consider such a scope if you need to use your rifle's longer range.

Less expensive and often quicker, but not quite as accurate, is learning to judge distances during practice by stepping them off and practicing making estimates. If you're fighting in a familiar area, you should also create a map which shows the previously stepped-off ranges to various landmarks. These can be used to make accurate estimates of an enemy's distance. In addition to practicing estimating and stepping off distances, it is also possible to estimate range by an enemy's apparent size. You'll have to practice this technique, but it can be quite accurate.

Although you can practically ignore windage at ranges under 100 yards, beyond that distance windage can make enough difference to affect whether or not you hit your target. To compensate for wind deflection, you must first guesstimate the wind speed. To do this, note the angle of deflection the wind creates on some light object. If a cloth is blowing in the wind, notice how high it is from what it's hanging on. The angle formed between the cloth and its support is divided by four to determine the wind speed. If a lightweight material isn't around, you can, in theory, create your own. Place your back to the wind and drop a cloth or piece of paper from shoulder height from a position where it would hit your foot in still conditions. Note where the object falls. Point at where it lands and estimate the angle created between your arm and body. Divide the angle by four and you'll obtain the wind speed.

Many times, however, you won't want to stand up in combat to figure out the wind speed. A more practical method is to study the wind's effect on the environment. Here's the common military method, which is surprisingly accurate: a wind under three mph can hardly be felt but will cause smoke to drift; three- to five-mph breezes can be lightly felt on the face; five- to eight-mph will keep tree leaves in constant motion; 8- to 12-mph will raise dust or blow loose papers; and a 12- to 15-mph wind will cause small trees to sway.

If the wind is blowing from your side, windage adjustment is pretty simple provided you've studied the wind-deflection ballistics of your round (see Appendix C). Adjust windage according to the chart, or if you're in a hurry, aim to one side of the target. If you have time, adjust the windage knob of the rifle; be sure you know how far one click will move the point of impact. It is wise to mark the zero adjustment point on the windage drum with a dab of paint so that you can rezero the weapon quickly.

If the wind is blowing from in front or in back of you, make no compensation. If the wind is coming toward your front or back at an oblique angle, give it a fractional value (roughly one-third to two-thirds of the full side-wind value).

When an enemy is moving, you must aim at where he'll be when the bullet reaches him. Here are the rules:

1. When a foe is moving parallel to you, your point of aim is the front edge of his body if he is within 250 yards.
2. If he is moving parallel to you beyond 250 yards, aim one body width in front of him.
3. An enemy running parallel to you should have a double lead, i.e., double his body width beyond 250 yards and one body width below 250 yards.
4. An enemy moving diagonally from you should have half the lead that you allowed for parallel movement.

It is also possible to shoot at an enemy running from one location to another almost as if he were stationary. Wait until he pauses, and fire.

At close ranges you may not have time to aim. This is especially true in urban or jungle combat. In such cases a "quick-kill" type of fire will improve your chances, as will multiple-projectile rounds or short bursts of automatic fire. Fire from the shoulder if possible and use a standard stance, but look over the sights of your weapon so that the barrel is parallel to where you're looking. Like everything else, this only works if you practice a lot.

Full training techniques are given in the U.S. Army's training manual 23-71-1, *Principles of Quick Kill* (available from Paladin Press). This style of shooting is easy to learn and can be essential in combat.

Success in combat is often a result of luck, but you can improve the odds in your favor by practicing a lot and remembering how your ammunition will behave within the various ranges at which you may be using it and how much barrier penetration it may offer.

Good ammunition and special loads can also improve your chances. Give some thought to your needs and select your ammunition carefully.

Appendix A: Manufacturers

A lot of different companies produce ammunition, reloading equipment, and reloading components, as well as useful tools and publications. Here's a list of a few which I've found to give good service.

AMMUNITION AND COMPONENTS

American Ballistics Co.
P.O. Box 1410
Marietta, GA 30061
(Subsonic ammunition for sound suppressor use, armor-piercing ammunition)

Barnes Bullets
P.O. Box 215
American Fork, UT 84003
(Bullets for reloaders)

Check-Mate Arms Co., Inc.
1421 Tower Square
Ventura, CA 93003
(9mm Geco-BAT ammunition)

DuPont
Explosives Dept.
Wilmington, DE 19898
(Powders for reloading, *Handloader's Guide)*

Euclid Sales Co.
1145 Euclid Ave. N.E.
Atlanta, GA 30307
(National brand tracers and exploding-bullet ammunition)

Federal Cartridge Company
2700 Foshay Tower
Minneapolis, MN 55402
(Combat ammunition for pistols and rifles)

Glaser Safety Slug
711 Somerset Lane, Box 8223
Forest City, CA 94404
(Glaser Safety Slugs)

Hercules Inc.
Hercules Plaza
Wilmington, DE 19894
(Reloading powders)

Hodgdon Powder Co., Inc.
P.O. Box 2805
Shawnee Mission, KS 66201
(Reloading powders, *Reloading Data Manual)*

Hornady Bullets
P.O. Box 1848
Grand Island, NE 68802
(Bullets for reloaders, *Hornady Handbook of Cartridge Reloading)*

Norma/Outdoor Sports HQ, Inc.
967 Watertower Ln.
Dayton, OH 45449
(Ammunition)

Patton and Morgan Corp.
5900 Wilshire Blvd., Suite 1400
Los Angeles, CA 80036
(Ammunition)

Phoenix Systems
P.O. Box 3339
Evergreen, CO 80439
(Grenade-launching cartridges, practice grenades)

Remington Arms Co.
939 Barnum Ave.
P.O. Box 1939
Bridgeport, CT 06601
(Ammunition)

Sierra Bullets
10532 South Painter Ave.
Santa Fe Springs, CA 90670
(Bullets for reloaders, including "MatchKing" bullets, *Reloader's Manual)*

Speer, c/o Omark Industries
P.O. Box 856
Lewiston, ID 83501
(Bullets, other reloading components)

Super-Vel, FPC, Inc.
Hamilton Rd., Rt. 2
Fond du Lac, WI 54935
(Super-Vel pistol ammunition)

Winchester Western (Olin)
120 Long Ridge Rd.
Stamford, CT 06904
(Ammunition)

RELOADING EQUIPMENT MANUFACTURERS

B-Square Company
P.O. Box 11281
Fort Worth, TX 76110
(Calipers, other tools)

Bonanza Sports, Inc.
412 Western Ave.
Faribault, MN 55021
(Reloading dies, scales, etc.)

C-H Tool & Die Corp.
106 N. Harding St.
Owen, WI 54460
(Reloading presses, dies, and other supplies, swaging dies for pistol-caliber bullets)

Corbin
P.O. Box 2659
White City, OR 97503
(Swaging dies, .22 LR brass-to-.224 jacket dies)

Dillon Precision Products, Inc.
7442 E. Butherus Dr.
Scottsdale, AZ 85250
(Reloading presses)

Forster Products
82 E. Lanark Ave.
Lanark, IL 61046
(Reloading equipment, tools)

I.S.W.
106 E. Cairo Dr.
Tempe, AZ 85282
(Case neck turner)

Lee Precision, Inc.
4275 Hwy. U
Hartford, WI 53027
(Presses and dies for most rifle and pistol calibers, inexpensive shotgun presses)

Lyman Products
Rte. 147
Middlefield, CT 06455
(Presses and reloading supplies, bullet-casting equipment)

Mequon Reloading Corp.
P.O. Box 253
Mequon, WI 53092
("Unitized" loader, reloading tools)

Monadnock Machine
Box 315, RR 2
Peterborough, NH 03458
(Brass trimmer)

Pacific Tool Co. (Hornady)
P.O. Drawer 2048, Ordnance Plant Rd.
Grand Island, NE 68801
(Presses, dies, related reloading equipment)

Quinetics Corp.
Box 28007
San Antonio, TX 78229
(Powder measures, kinetic bullet puller)

RCBS, c/o Omark Industries
P.O. Box 856
Lewiston, ID 83501
(Presses, dies, powder scales, dial calipers, etc., and a wide range of reloading supplies for the reloader)

Redding, Inc.
114 Starr Rd.
Cortland, NY 13045
(Brass trimmer)

ACCESSORIES FOR THE COMBAT AMMUNITION RELOADER

E & L Manufacturing
Star Route 2, Box 569
Cave Creek, AZ 85331
(Brass catchers for popular combat rifles and carbines)

Harry Owen
P.O. Box 5337
Hacienda Heights, CA 91745
(Cartridge adapters for combat rifles and shotguns)

PUBLICATIONS AND VIDEO TAPES

Anite Company
P.O. Box 375
Pinole, CA 94564
(Tapes demonstrating the abilities of various types of combat ammunition)

DBI Books
One Northfield Plaza
Northfield, IL 60093
(Reloading books, annual editions of *Gun Digest*)

Paladin Press
P.O. Box 1307
Boulder, CO 80306
(Books on firearms, combat, survival, related subjects)

RK Enterprises, Video Production Division
2616 Las Positas
Santa Barbara, CA 93105
(Tapes of pistol- and rifle-ammunition reloading techniques)

Appendix B: Ballistics Tables

Careful study of the following ballistic charts and those in reloading manuals and elsewhere will greatly aid you in getting your bullets to the target. Actual figures for your gun and ammunition may vary greatly from those shown below. These tables should be used for reference only.

PISTOL CALIBERS

Foot pounds of energy (fpe) are useful for comparing energy levels within a given caliber but do not indicate the wounding capability of different calibers. Factors such as bullet design, speed, etc., come into play, so that it is a mistake to assume that one caliber of bullet is more lethal than another because it has more foot-pounds of energy at a given distance.

Speeds are given in feet per second (fps) and energy levels in foot-pounds of energy (fpe).

.22 LONG RIFLE (40-GRAIN)

	Muzzle	50 yds.	100 yds.
Velocity (fps)	1200	1069	1016
Energy (fpe)	128	102	92

.22 LONG RIFLE HIGH-VELOCITY (32-GRAIN)

	Muzzle	50 yds.	100 yds.
Velocity (fps)	1560	1282	1090
Energy (fpe)	173	117	84

.25 AUTO (45-GRAIN)

	Muzzle	50 yds.	100 yds.
Velocity (fps)	815	729	655
Energy (fpe)	66	53	42

.32 AUTO (60-GRAIN)

	Muzzle	50 yds.	100 yds.
Velocity (fps)	970	895	835
Energy (fpe)	125	107	93

.380 AUTO (85-GRAIN)

	Muzzle	50 yds.	100 yds.
Velocity (fps)	1000	921	860
Energy (fpe)	189	160	140

.38 SPECIAL (110-GRAIN)

	Muzzle	50 yds.	100 yds.
Velocity (fps)	945	894	850
Energy (fpe)	218	195	176

.38 SPECIAL +P (95-GRAIN)

	Muzzle	50 yds.	100 yds.
Velocity (fps)	1100	1002	932
Energy (fpe)	255	212	183

9MM LUGER (95-GRAIN)

	Muzzle	50 yds.	100 yds.
Velocity (fps)	1355	1140	1008
Energy (fpe)	387	274	214

9MM LUGER (115-GRAIN)

	Muzzle	50 yds.	100 yds.
Velocity (fps)	1225	1095	1007
Energy (fpe)	383	306	259

.38 SUPER AUTO +P (125-GRAIN)

	Muzzle	50 yds.	100 yds.
Velocity (fps)	1240	1130	1050
Energy (fpe)	427	354	306

.38 SUPER AUTO +P (130-GRAIN)

	Muzzle	50 yds.	100 yds.
Velocity (fps)	1215	1099	1017
Energy (fpe)	426	348	298

.357 MAGNUM (110-GRAIN)

	Muzzle	50 yds.	100 yds.
Velocity (fps)	1295	1095	975
Energy (fpe)	410	292	232

.357 MAGNUM (145-GRAIN)

	Muzzle	50 yds.	100 yds.
Velocity (fps)	1290	1155	1060
Energy (fpe)	535	428	361

.41 MAGNUM (175-GRAIN)

	Muzzle	50 yds.	100 yds.
Velocity (fps)	1250	1120	1029
Energy (fpe)	607	488	412

.41 MAGNUM (210-GRAIN)

	Muzzle	50 yds.	100 yds.
Velocity (fps)	1300	1182	1062
Energy (fpe)	788	630	526

.44 MAGNUM (210-GRAIN)

	Muzzle	50 yds.	100 yds.
Velocity (fps)	1250	1106	1010
Energy (fpe)	729	570	475

.44 MAGNUM (240-GRAIN)

	Muzzle	50 yds.	100 yds.
Velocity (fps)	1350	1186	1069
Energy (fpe)	971	749	608

.45 AUTO (185-GRAIN)

	Muzzle	50 yds.	100 yds.
Velocity (fps)	1000	938	888
Energy (fpe)	411	362	324

.45 AUTO (230-GRAIN)

	Muzzle	50 yds.	100 yds.
Velocity (fps)	810	776	745
Energy (fpe)	335	308	284

RIFLE CALIBERS

When comparing cartridges and loads, remember that a bullet's high-velocity wounding capability requires a velocity of 2000 fps. If the bullet drops below this speed, it will lose a lot of its potential regardless of caliber. This drop can be created by distance or barrel length; most of the figures shown below and in other ballistic tables apply to standard-length barrels; if your rifle has a short barrel, take that into account.

With bullets traveling in the 2000- to 3000-fps range, one inch of barrel length will make around a 20 fps velocity change. A 10-inch barrel will have a muzzle velocity about 200 fps below its 20-inch counterpart. While this may not make much difference in close combat, at ranges of 250 plus yards, it may critically affect your rifle's lethality.

As with pistol cartridges, fpe are useful for comparing the energy levels within a specific caliber but do not tell a lot about the wounding capability of different calibers of bullets.

.223 REMINGTON (5.56mm)
55-GRAIN (FMJ) .224 WITH 250-YARD ZERO

	Muzzle	100 yds.	200 yds.	300 yds.	400 yds.	500 yds.	600 yds.
Div from 0	-1.5"	1.5"	2.5"	-3"	-17.8"	-45"	-90"
Velocity (fps)	3100	2640	2226	1859	1544	1277	1093
Energy (fpe)	1174	852	605	422	291	199	146

.223 REMINGTON (5.56mm)
55-GRAIN (FMJ) .224 WITH 250-YARD ZERO (MAXIMUM LOAD)

	Muzzle	100 yds.	200 yds.	300 yds.	400 yds.	500 yds.	600 yds.
Div from 0	-1.5"	1"	2.8"	-4.1"	-18.3"	-41.6"	-87"
Velocity (fps)	3240	2877	2543	2232	1943	1679	1455
Energy (fpe)	1282	1011	790	608	461	344	259

.223 REMINGTON (5.56mm)
55-GRAIN (FMJ) .224 WITH 300-YARD ZERO

	Muzzle	100 yds.	200 yds.	300 yds.	400 yds.	500 yds.	600 yds.
Div from 0	-1.5"	4.8"	6"	0"	-16.4"	-47"	-98"
Velocity (fps)	3100	2640	2226	1859	1544	1277	1093
Energy (fpe)	1174	852	605	422	291	199	146

.223 REMINGTON (5.56mm)
55-GRAIN (FMJ) .224 WITH 300-YARD ZERO (MAXIMUM LOAD)

	Muzzle	100 yds.	200 yds.	300 yds.	400 yds.	500 yds.	600 yds.
Div from 0	-1.5"	4.4"	5.6"	0"	-15.1"	-43.8"	-91.7"
Velocity (fps)	3200	2732	2307	1931	1604	1324	1126
Energy (fpe)	1251	912	650	456	314	214	155

.223 REMINGTON (5.56mm)
60-GRAIN .224 WITH 250-YARD ZERO

	Muzzle	100 yds.	200 yds.	300 yds.	400 yds.	500 yds.	600 yds.
Div from 0	-1.5"	3.25"	2.9"	-4.3"	-20"	-48"	-92.5"
Velocity (fps)	3000	2633	2296	1991	1716	1467	1260
Energy (fpe)	1199	924	703	528	392	287	211

.223 REMINGTON (5.56mm)
60-GRAIN .224 WITH 250-YARD ZERO (MAXIMUM LOAD)

	Muzzle	100 yds.	200 yds.	300 yds.	400 yds.	500 yds.	600 yds.
Div from 0	-1.5"	1.9"	2.9"	-2"	-14.8"	-37.6"	-74"
Velocity (fps)	3200	2820	2467	2145	1854	1596	1365
Energy (fpe)	1365	1060	811	613	458	340	248

.223 REMINGTON (5.56mm)
69-GRAIN .224 WITH 250-YARD ZERO (MAXIMUM LOAD)

	Muzzle	100 yds.	200 yds.	300 yds.	400 yds.	500 yds.	600 yds.
Div from 0	-1.5"	2"	3.5"	-4"	-12.4"	-35"	-71"
Velocity (fps)	3000	2724	2462	2214	2004	1786	1555
Energy (fpe)	1379	1137	929	751	615	488	370

.30 CARBINE
110-GRAIN .308 WITH 100-YARD ZERO

	Muzzle	100 yds.	200 yds.	300 yds.	400 yds.	500 yds.
Div from 0	0.5"	0.0"	-13.5"	-49.9"	-118.6"	-228.2"
Velocity (fps)	1990	1567	1236	1035	923	842
Energy (fpe)	967	600	373	262	208	173

7.62X39mm RUSSIAN
110-GRAIN .308 WITH 250-YARD ZERO

	Muzzle	100 yds.	200 yds.	300 yds.	400 yds.
Div from 0	-1.5"	6.9"	6.0"	-11"	-46.7"
Velocity (fps)	2500	1956	1500	1162	970
Energy (fpe)	1527	935	550	330	230

7.62X39mm RUSSIAN
150-GRAIN .308 WITH 300-YARD ZERO

	Muzzle	100 yds.	200 yds.	300 yds.	400 yds.	500 yds.
Div from 0	-1.5"	8.7"	9.9"	0"	-23.6"	-65"
Velocity (fps)	2200	1980	1777	1592	1417	1266
Energy (fpe)	1612	1306	1051	844	669	534

.308 WINCHESTER (7.62mm NATO)
150-GRAIN .308 WITH 250-YARD ZERO

	Muzzle	100 yds.	200 yds.	300 yds.	400 yds.	500 yds.	600 yds.
Div from 0	-1.5"	3.5"	3"	-4.5"	-20"	-47"	-90"
Velocity (fps)	2820	2593	2396	2210	2035	1869	1714
Energy (fpe)	2648	2240	1913	1628	1379	1164	979

.308 WINCHESTER (7.62mm NATO)
165-GRAIN .308 WITH 300-YARD ZERO

	Muzzle	100 yds.	200 yds.	300 yds.	400 yds.	500 yds.	600 yds.
Div from 0	-1.5"	5.8"	6.8"	0"	-15.8"	-42.8"	-82.8"
Velocity (fps)	2600	2378	2169	1974	1791	1623	1462
Energy (fpe)	2477	2072	1724	1427	1175	966	783

.308 WINCHESTER (7.62mm NATO)
180-GRAIN .308 WITH 250-YARD ZERO

	Muzzle	100 yds.	200 yds.	300 yds.	400 yds.	500 yds.	600 yds.
Div from 0	-1.5"	4"	3.25"	-4.8"	-21.75"	-49.7"	-90"
Velocity (fps)	2600	2393	2198	2015	1842	1682	1535
Energy (fpe)	2703	2290	1932	1623	1357	1131	942

.30-06
150-GRAIN .308 WITH 300-YARD ZERO

	Muzzle	100 yds.	200 yds.	300 yds.	400 yds.	500 yds.	600 yds.
Div from 0	-1.5"	5.1"	5.9"	0"	-14.0"	-38.0"	-73.4"
Velocity (fps)	3000	2729	2473	2234	2012	1806	1618
Energy (fpe)	2998	2481	2038	1663	1349	1087	873

.30-06
180-GRAIN .308 WITH 300-YARD ZERO

	Muzzle	100 yds.	200 yds.	300 yds.	400 yds.	500 yds.	600 yds.
Div from 0	-1.5"	5.1"	5.9"	0"	-14.0"	-38.0"	-73.4"
Velocity (fps)	2700	2488	2287	2098	1921	1754	1602
Energy (fpe)	2914	2474	2091	1760	1475	1230	1025

SHOTGUN

Shotgun slugs vary a great deal in weight and performance. Barrel length and choke type will also affect accuracy and velocity. Nevertheless, the following chart should give you some idea of what a 12-gauge slug can—and can't—do.

12-GAUGE SLUG
1 OUNCE WITH 75-YARD ZERO

	Muzzle	25 yds.	50 yds.	75 yds.	100 yds.	125 yds.	150 yds.
Div from 0	–0.5"	0.5"	1.0"	0"	-3.0"	-8.1"	-15.7"
Velocity (fps)	1560	1345	1175	1057	977	916	865
Energy (fpe)	2365	1756	1341	1085	926	814	727

Appendix C: Wind-Deflection Tables

To really learn about wind deflection, a lot of practice under bad conditions is necessary. It can be a real eye-opener to see a bullet kick up dust a foot away from where you were aiming! For wind speeds other than those shown, divide the speed by 10 and multiply the number times the deflection shown for the 10 mph column. For example, the 3 mph wind deflection for a 55-grain .223 Remington bullet with a muzzle velocity of 3240 fps at 200 yards would be: 3/10 x 4.4 = 1.32-inch deflection.

WIND DEFLECTION ON 115-GRAIN 9mm LUGER (1150 FPS MUZZLE VELOCITY)

Mph Wind	Muzzle	50 yds.	100 yds.	150 yds.	200 yds.	250 yds.	300 yds.
4	0	.44"	1.7"	3.8"	6.7"	10"	16"
8	0	.88"	3.4"	7.6"	13.4"	20"	32"
10	0	1.1"	4.3"	9.5"	16.7"	25"	38"
20	0	2.2"	8.6"	19"	33.4"	50"	76"

WIND DEFLECTION ON 185-GRAIN .45 AUTO (850 FPS MUZZLE VELOCITY)

Mph Wind	Muzzle	50 yds.	100 yds.	150 yds.	200 yds.	250 yds.	300 yds.
4	0	.4"	1.8"	4.2"	7.2"	11.4"	17.2"
8	0	.8"	3.7"	8.3"	14.5"	22.9"	34.3"
10	0	1"	4.6"	10.4"	18.1"	28.6"	42.9"
20	0	2"	9.2"	20.8"	36.2"	57.2"	85.8"

WIND DEFLECTION ON 55-GRAIN .223 REMINGTON BULLET (3240 FPS MUZZLE VELOCITY)

Mph Wind	Muzzle	100 yds.	200 yds.	300 yds.	400 yds.	500 yds.	600 yds.
4	0	.44"	1.6"	4"	8"	13.6"	20"
8	0	.88"	3.5"	7.9"	16"	27.2"	40"
10	0	1"	4.4"	10.6"	20.1"	34"	53.3"
20	0	2.2"	8.9"	20"	40.2"	68"	100"

WIND DEFLECTION ON 69-GRAIN .223 REMINGTON BULLET (3000 FPS MUZZLE VELOCITY)

Mph Wind	Muzzle	100 yds.	200 yds.	300 yds.	400 yds.	500 yds.	600 yds.
4	0	.4"	1.5"	2.6"	5.7"	10.7"	15.5"
8	0	.8"	2.9"	5.2"	11.3"	21.4"	31"
10	0	.9"	3.7"	7"	14.2"	27.9"	41"
20	0	2"	7.4"	14"	28.1"	56"	82"

WIND DEFLECTION ON 110-GRAIN (.308) CARBINE (1990 FPS MUZZLE VELOCITY)

Mph Wind	Muzzle	100 yds.	200 yds.	300 yds.	400 yds.	500 yds.	600 yds.	700 yds.
4	0	0.9"	3.9"	10"	16.8"	26.4"	39"	56"
8	0	1.8"	7.8"	19"	33.6"	52.8"	78"	112"
10	0	2.2"	9.7"	24"	42"	66"	98"	140"
20	0	4.4"	19.4"	48"	84"	132"	196"	280"

WIND DEFLECTION ON 110-GRAIN (.308) 7.62X39mm RUSSIAN (2500 FPS MUZZLE VELOCITY)

Mph Wind	Muzzle	100 yds.	200 yds.	300 yds.	400 yds.	500 yds.	600 yds.	700 yds.
4	0	.6"	2.9"	7"	13.3"	22.1"	34"	50"
8	0	1.3"	5.8"	14"	26.7"	44.2"	67"	100"
10	0	1.6"	7.2"	17.5"	33.4"	55.3"	84"	125"
20	0	3.2"	14.4"	35"	66.8"	110.6"	168"	250"

WIND DEFLECTION ON 150-GRAIN (.308) 7.62X39mm RUSSIAN (2200 FPS MUZZLE VELOCITY)

Mph Wind	Muzzle	100 yds.	200 yds.	300 yds.	400 yds.	500 yds.	600 yds.	700 yds.
4	0	0.4"	1.7"	4"	7.5"	12.3"	19"	28"
8	0	0.8"	3.4"	8"	15"	24.6"	37"	56"
10	0	1.0"	4.3"	10.1"	18.8"	30.7"	47"	70"
20	0	2.0"	8.6"	20.2"	37.6"	61.4"	94"	140"

WIND DEFLECTION ON 150-GRAIN .308 WINCHESTER/.30-06 BULLET (3000 FPS MUZZLE VELOCITY)

Mph Wind	Muzzle	100 yds.	200 yds.	300 yds.	400 yds.	500 yds.	600 yds.	700 yds.
4	0	.2"	1.1"	2.7"	5"	8.1"	13"	18"
8	0	.5"	2.2"	5.4"	9.9"	16.2"	26"	37"
10	0	.6"	2.8"	6.7"	12.4"	20.2"	32"	46"
20	0	1.2"	5.6"	13.4"	24.8"	40.4"	64"	92"

WIND DEFLECTION ON 180-GRAIN .308 WINCHESTER/.30-06 BULLET (2500 FPS MUZZLE VELOCITY)

Mph Wind	Muzzle	100 yds.	200 yds.	300 yds.	400 yds.	500 yds.	600 yds.	700 yds.
4	0	.4"	1.6"	3.8"	7.2"	12"	18"	26"
8	0	.7"	3.3"	7.7"	14.3"	23"	36"	53"
10	0	.9"	4.1"	9.6"	17.9"	29"	45"	66"
20	0	1.8"	8.2"	19.2"	35.8"	58"	90"	132"

Appendix D: Conversion Formulas

There are a number of formulas which I've found very useful over the years. I've included them here in the hope that they will serve you as well as they have me. I have included some English to Metric conversion formulas for those who aren't as enamored with the English system as I am.

OUNCES TO GRAINS CONVERSION

Ever wonder how many reloads you could get from a pound of powder? The secret is to convert the weight to ounces and then to grains and divide by the number of grains in your load. All you need to know is that there are 437.5 grains (gr.) per ounce (oz.) and 16 ounces in a pound (lb.).

FINDING A BULLET'S ENERGY

Determining the energy of a bullet is quite simple and is sometimes useful (though it is sometimes a bit misleading as well).

To find the energy of a bullet, square its velocity in feet per second and divide the result by 450,240. This will give you the energy per grain, so the last step is to multiply your result by the bullet's weight in grains. This will give you its energy in foot pounds. Here's the formula:

$$\text{Velocity}^2 / 450{,}240 \times \text{Weight} = \text{Energy}$$

For example, if we had a 90-grain bullet traveling at 1200 fps, its energy would be found as follows:

(1200 fps x 1200 fps) divided by 450,240 times 90 grains
1440000 / 450,240 x 90
3.1982942 x 90
287.84647 fpe

or, if we round it off: 288 fpe.

ENGLISH/METRIC CONVERSIONS

While the yard is roughly equivalent to the meter, roughly doesn't cut it sometimes. When you need to get really precise, here's how to do it.

English to metric (length):

Inches (in.) x 25.40005 = millimeters (mm)
Inches x 2.540005 = centimeters (cm)
Feet (ft.) x 0.3048007 = meters (m)
Yards (yd.) x 0.914402 = meters (m)
Miles x 1.6093419 = kilometers

English to metric (velocity):

Fps x 0.3048007 = mps (meters per second)

English to metric (energy):

Fpe x 0.1383 = kgm (kilogram-meter)

English to metric (weight):

Grains (gr.) x 0.0647989 = grams
Ounces (oz.) x 28.349527 = grams
Pounds (lbs.) x 0.4535924 = kilograms

English to metric (volume):

Cubic inches x 16.387156 = cubic centimeters

METRIC/ENGLISH CONVERSIONS

Metric to English (length):

Millimeters x 0.0393 = inches
Centimeters x 0.3937 = inches
Meters x 3.2808321 = feet
Meters x 1.0936109 = yards
Kilometers x 0.6213327 = miles

Metric to English (velocity):

Mps x 3.2808321 = fps

Metric to English (energy):

Kgm x 7.233 = fpe

Metric to English (weight):

Grams x 15.43236 = grains
Grams x 0.0352739 = ounces
Kilograms x 2.2046224 = pounds

Metric to English (volume):

Cubic centimeters x 0.0610234 = cubic inches

Appendix E: Ballistic Gelatin Results

.223/5.56mm SPITZER PERFORMANCE IN BALLISTIC GELATIN
55-Grain, Spitzer (Lead-Point) Bullet
Velocity: 3200 fps

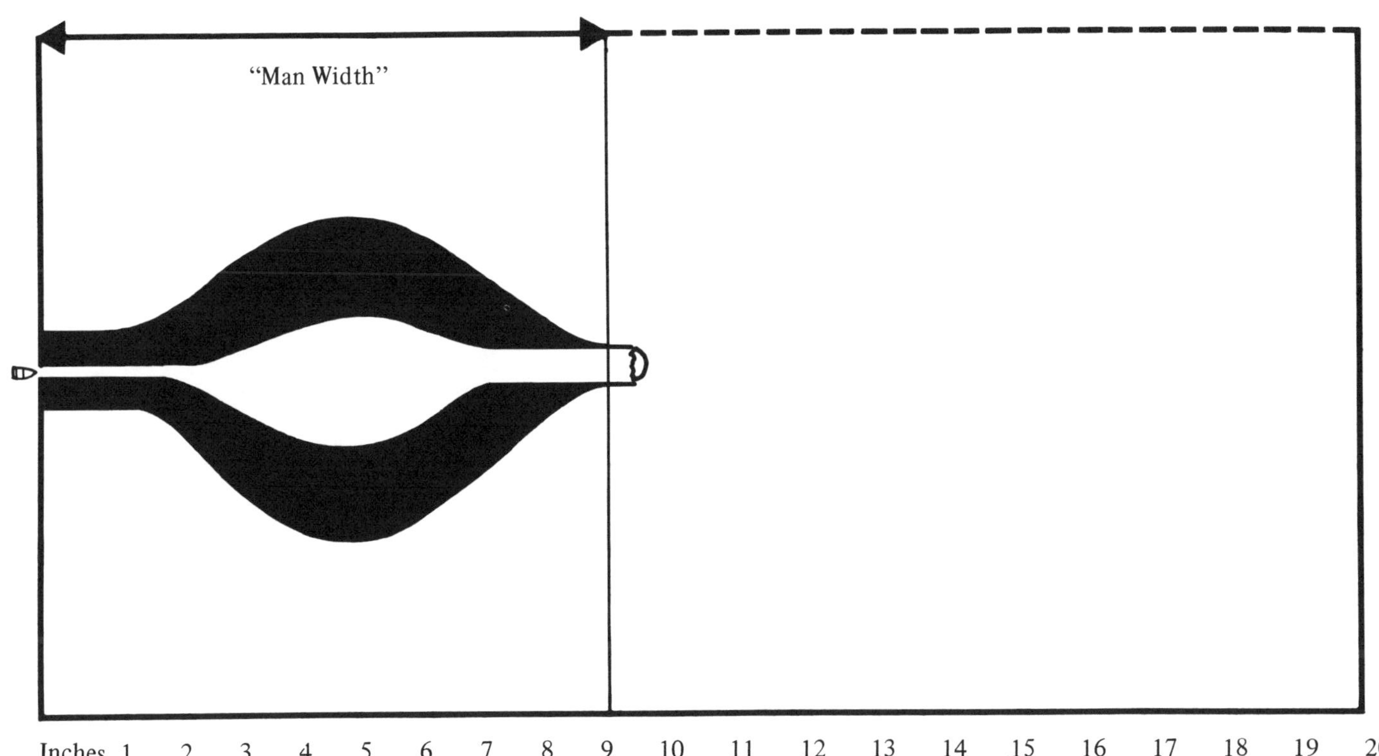

Permanent cavity moves down the center of the gelatin; the temporary cavity is shown in the shaded area, mostly within the 9-inch "man width." Final bullet diameter: 0.45-inch. An ideal combat round.

.223/5.56mm FMJ PERFORMANCE IN BALLISTIC GELATIN
55-Grain, FMJ Bullet
Velocity: 3200 fps; 1-in-12 Twist

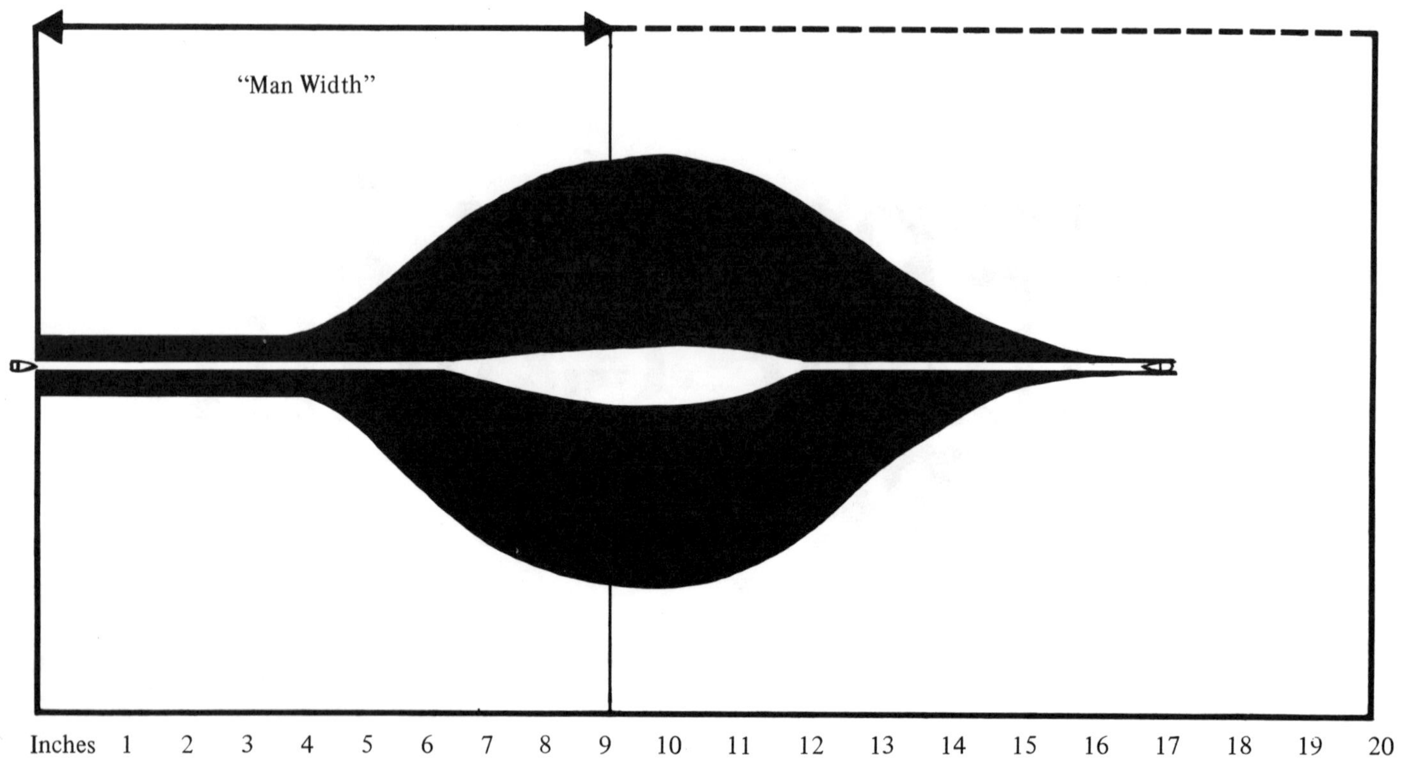

Permanent cavity moves down the center of the gelatin; the temporary cavity is shown in the shaded area, mostly beyond the 9-inch "man width." (This shows why the M16's 1-in-14 twist is more deadly than the 1-in-12.)

.223/5.56mm FMJ PERFORMANCE IN BALLISTIC GELATIN
55-Grain, FMJ Bullet
Velocity: 3200 fps; 1-in-14 Twist

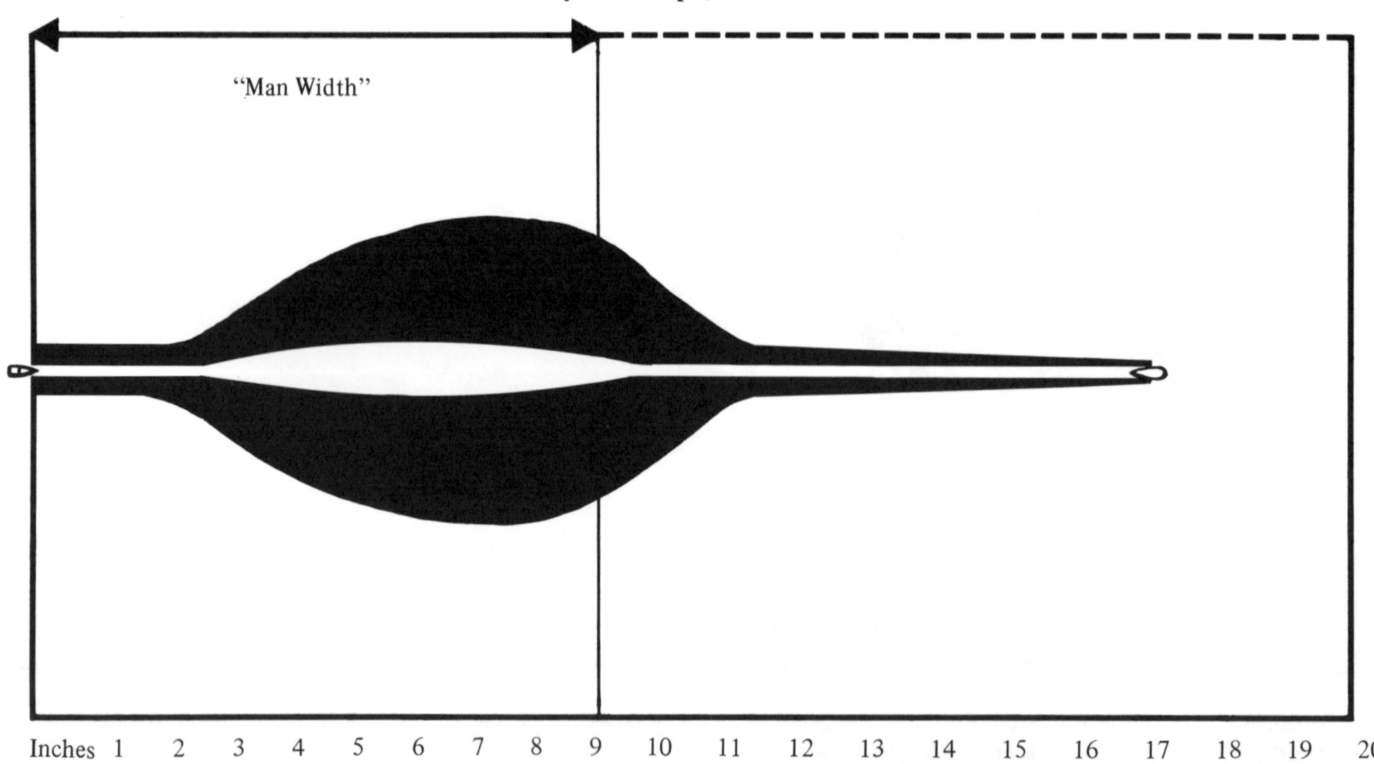

Permanent cavity moves down the center of the gelatin; the temporary cavity is shown in the shaded area, mostly within the 9-inch "man width." (Compare this to the 1-in-12 twist.)

.223/5.56mm PERFORMANCE IN BALLISTIC GELATIN
69-Grain, Jacketed Hollow-Point Bullet
Velocity: 2900 fps

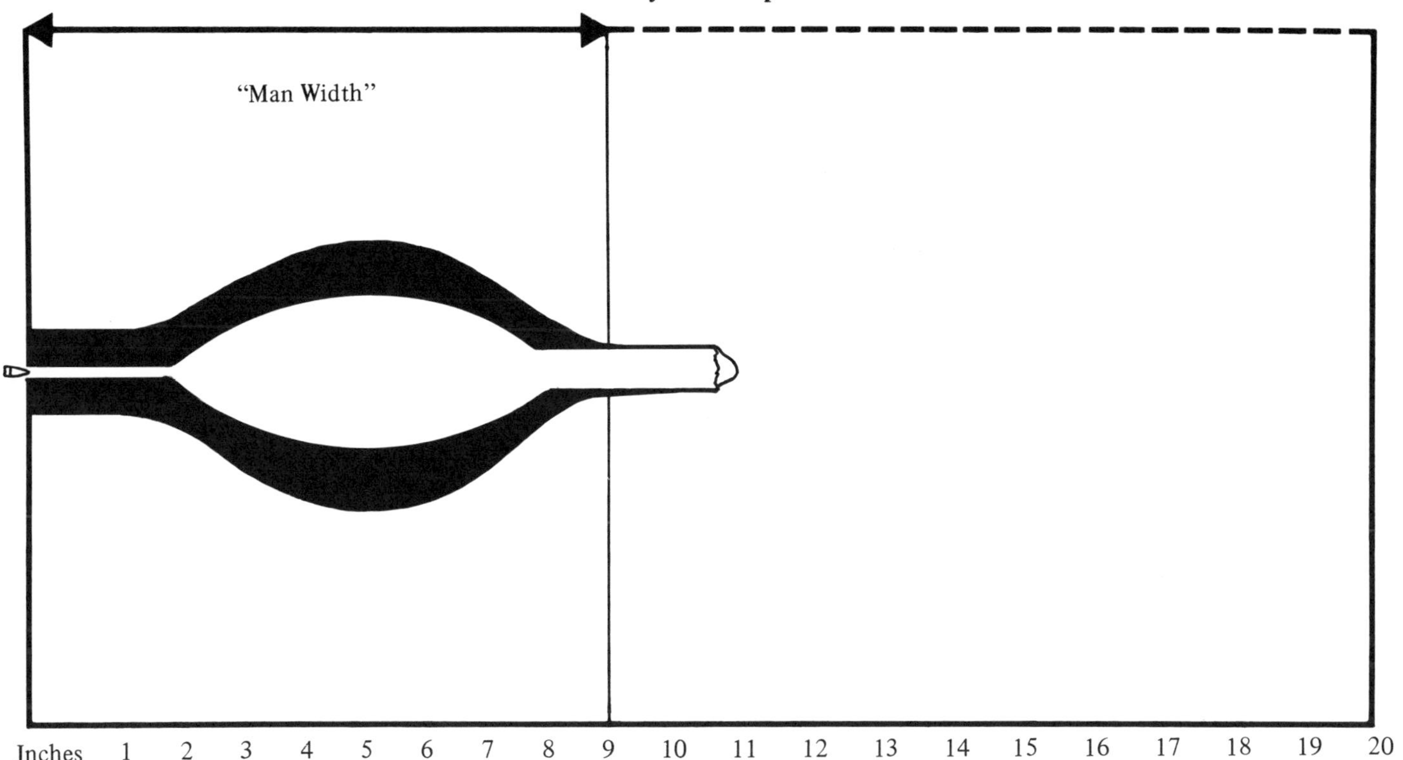

Permanent cavity moves down the center of the gelatin; the temporary cavity is shown in the shaded area within "man width." This bullet would be ideal for combat use.

7.62mm NATO FMJ PERFORMANCE IN BALLISTIC GELATIN
150-Grain, FMJ Bullet
Velocity: 2800 fps

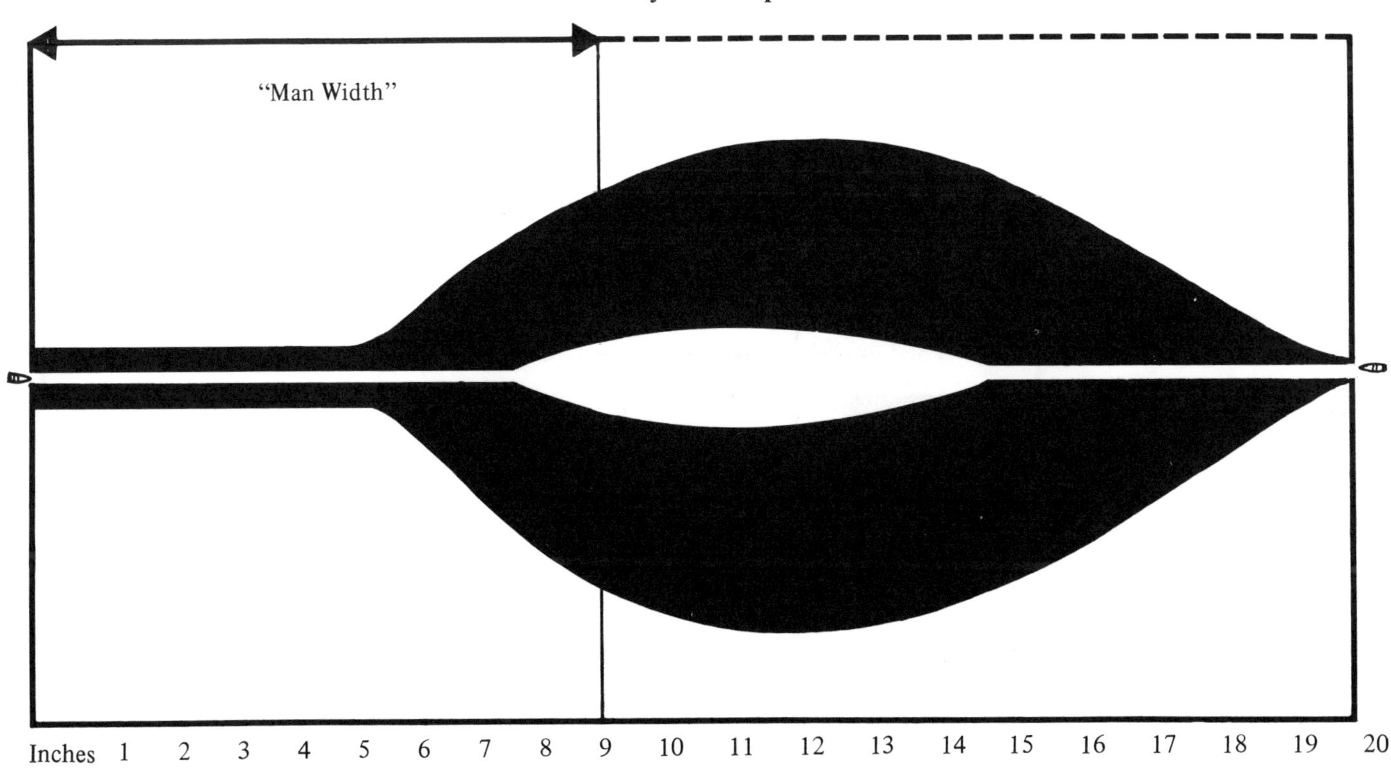

Permanent cavity moves down the center of the gelatin; the temporary cavity is shown in the shaded area. This bullet "tumbles" far too late for combat. (Note that maximum damage occurs after the 9-inch "man width.")

.308 WINCHESTER/.30-06 PERFORMANCE IN BALLISTIC GELATIN

150-Grain, Spitzer (Lead-Point) Bullet

Velocity: 2800 fps

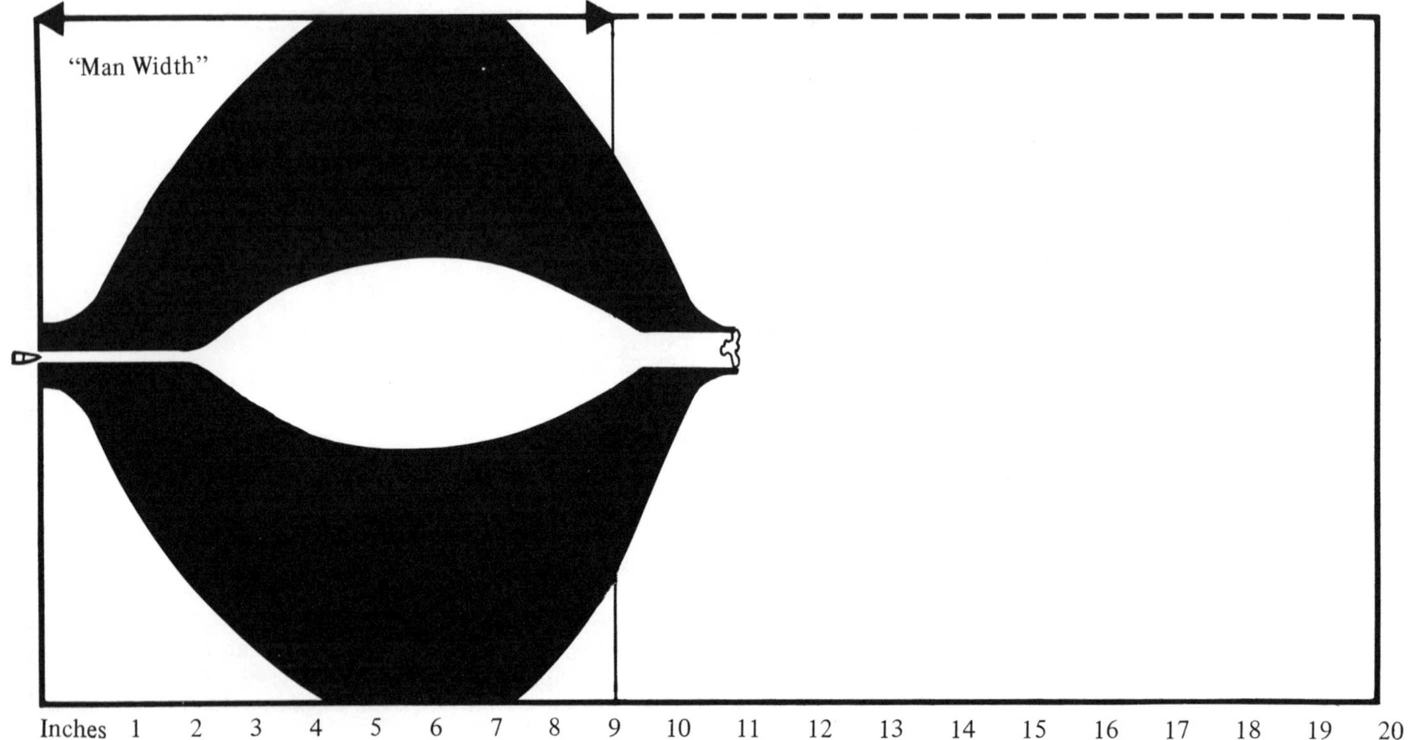

Permanent cavity moves down the center of the gelatin; the temporary cavity is shown in the shaded area within the 9-inch "man width." Unlike its FMJ counterpart, this bullet would be ideal for combat.

12-GAUGE SHOTGUN SLUG PERFORMANCE IN BALLISTIC GELATIN

437-Grain, Lead Bullet

Velocity: 1500 fps

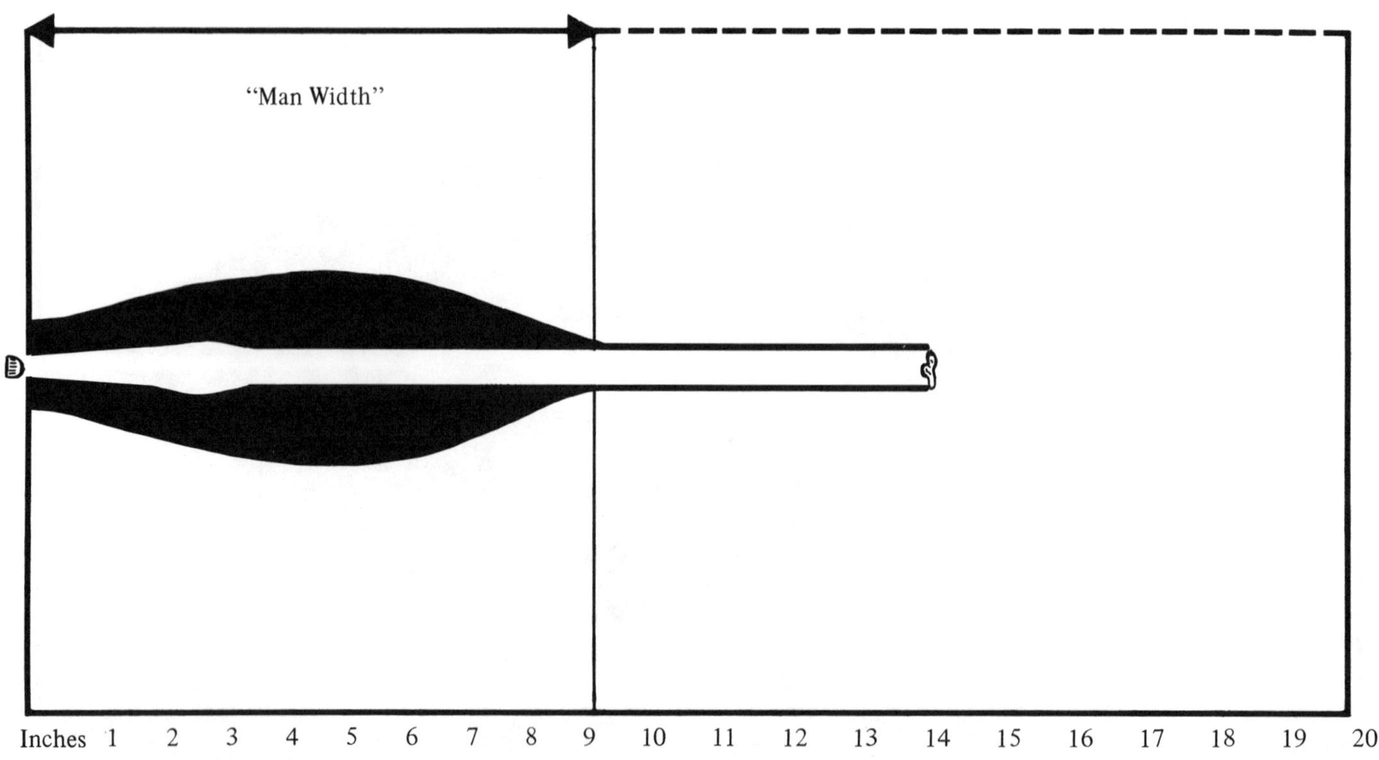

A permanent cavity is left in the gelatin (1.1-inch diameter); the temporary cavity is shown in the shaded area. From this illustration, we can see how the slug can be so devastating: maximum damage occurs in the 9-inch "man width."

.22 LR PERFORMANCE IN BALLISTIC GELATIN

40-Grain, Lead Bullet

Velocity: 1120 fps

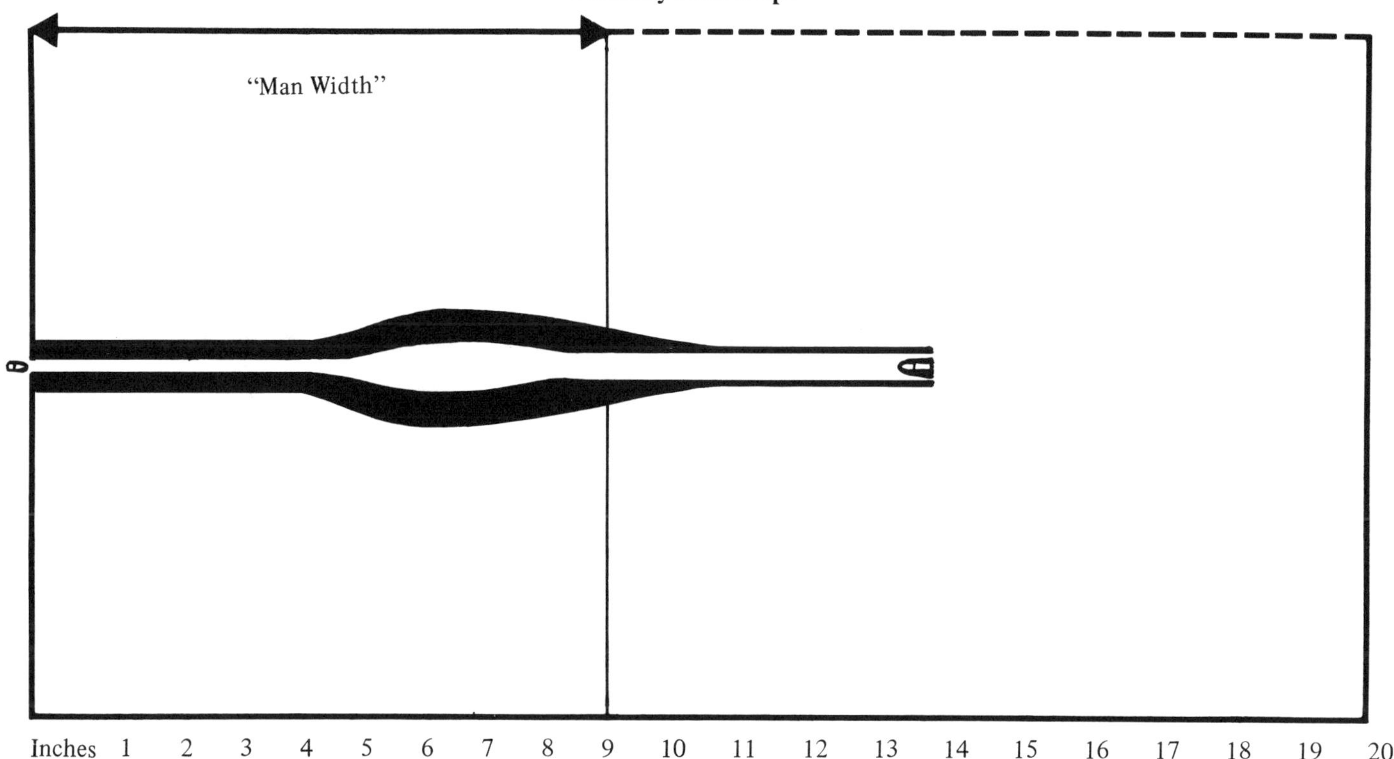

Permanent cavity moves down the center of the gelatin; the temporary cavity is shown in the shaded area within the 9-inch "man width." The bullet tumbles and often exits base first.

.22 LR PERFORMANCE IN BALLISTIC GELATIN

37-Grain, Hollow-Point Lead Bullet

Velocity: 1270 fps

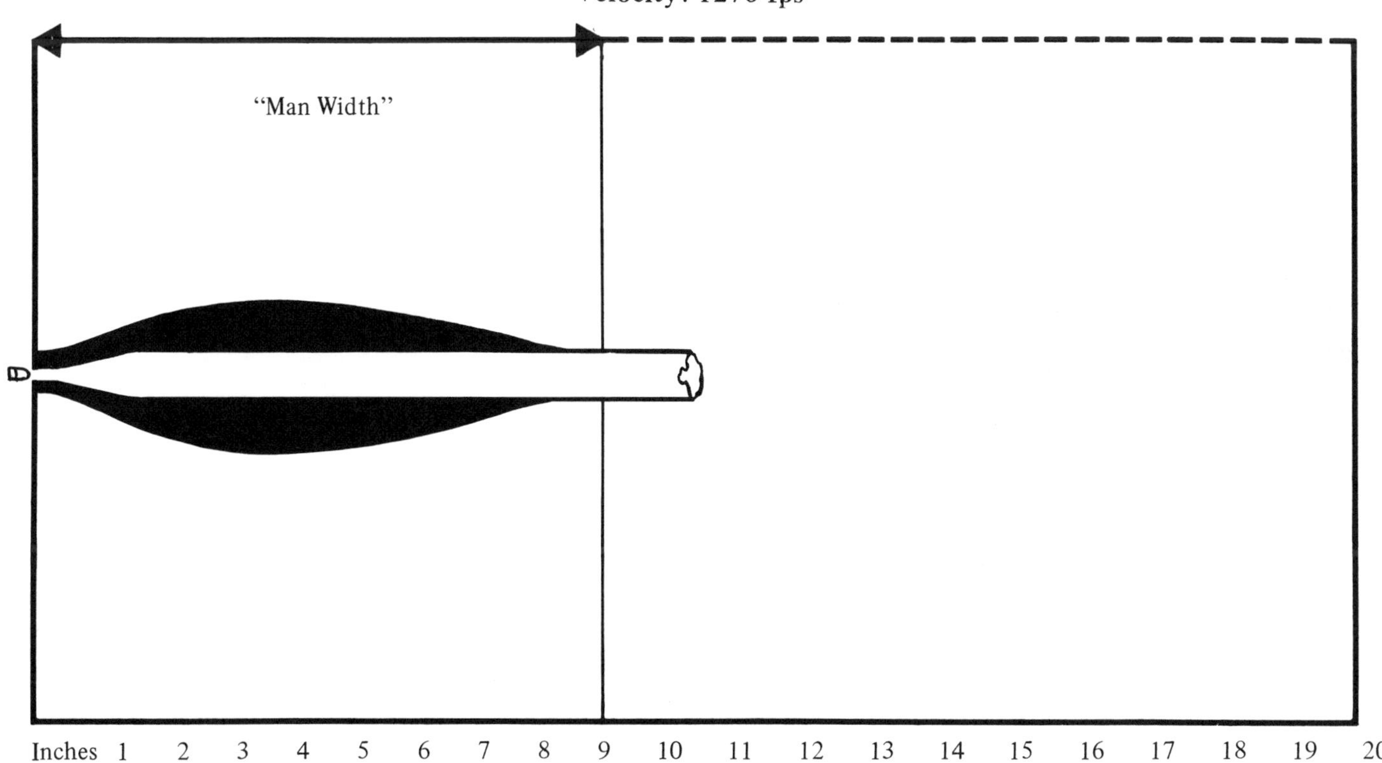

Permanent cavity moves down the center of the gelatin; the temporary cavity is shown in the shaded area within the 9-inch "man width." The bullet expands to .38-inch diameter.

.45 AUTO FMJ PERFORMANCE IN BALLISTIC GELATIN
230-Grain, FMJ Bullet
Velocity: 850 fps

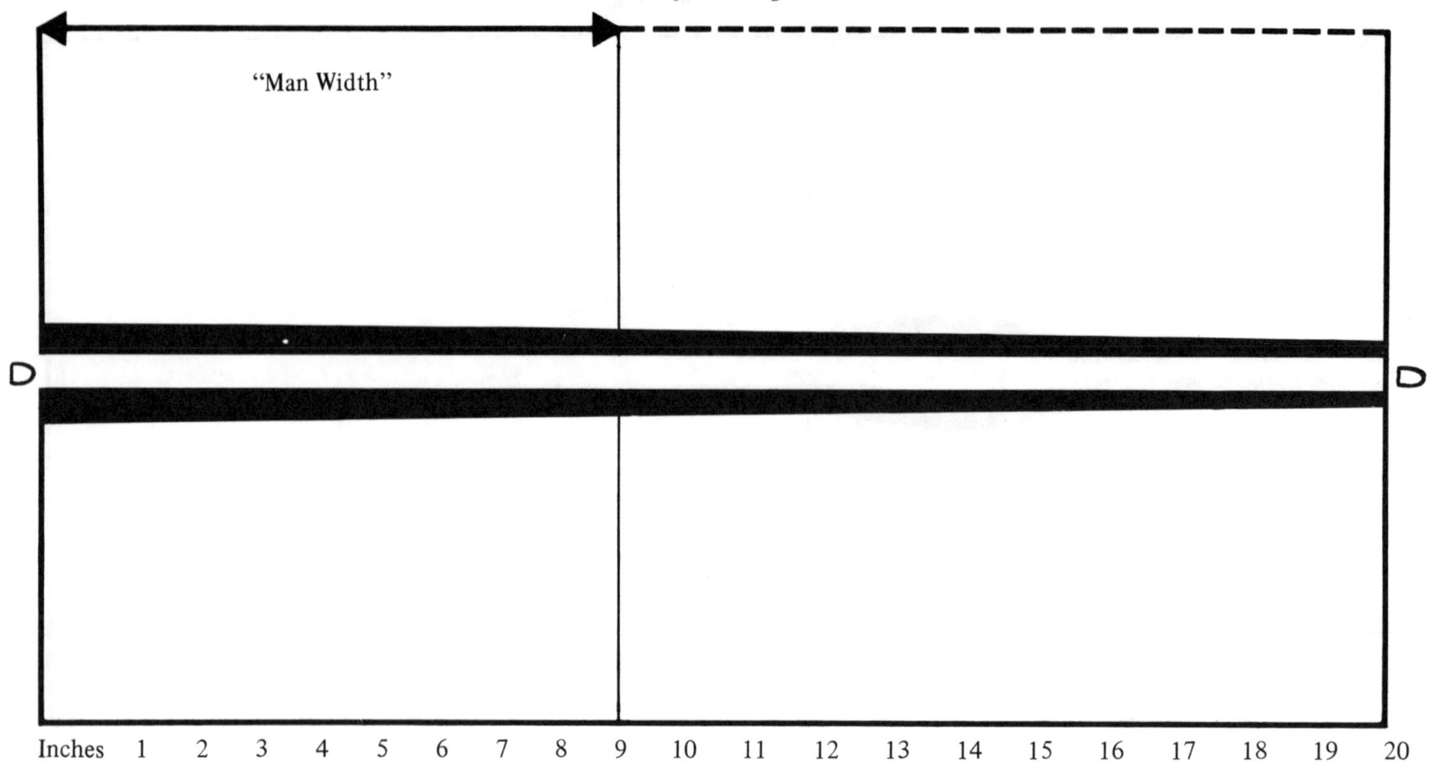

This bullet neither tumbles nor expands and is therefore far from ideal.

9mm LUGER PERFORMANCE IN BALLISTIC GELATIN
115-Grain, Jacketed Hollow-Point Bullet
Velocity: 1200 fps

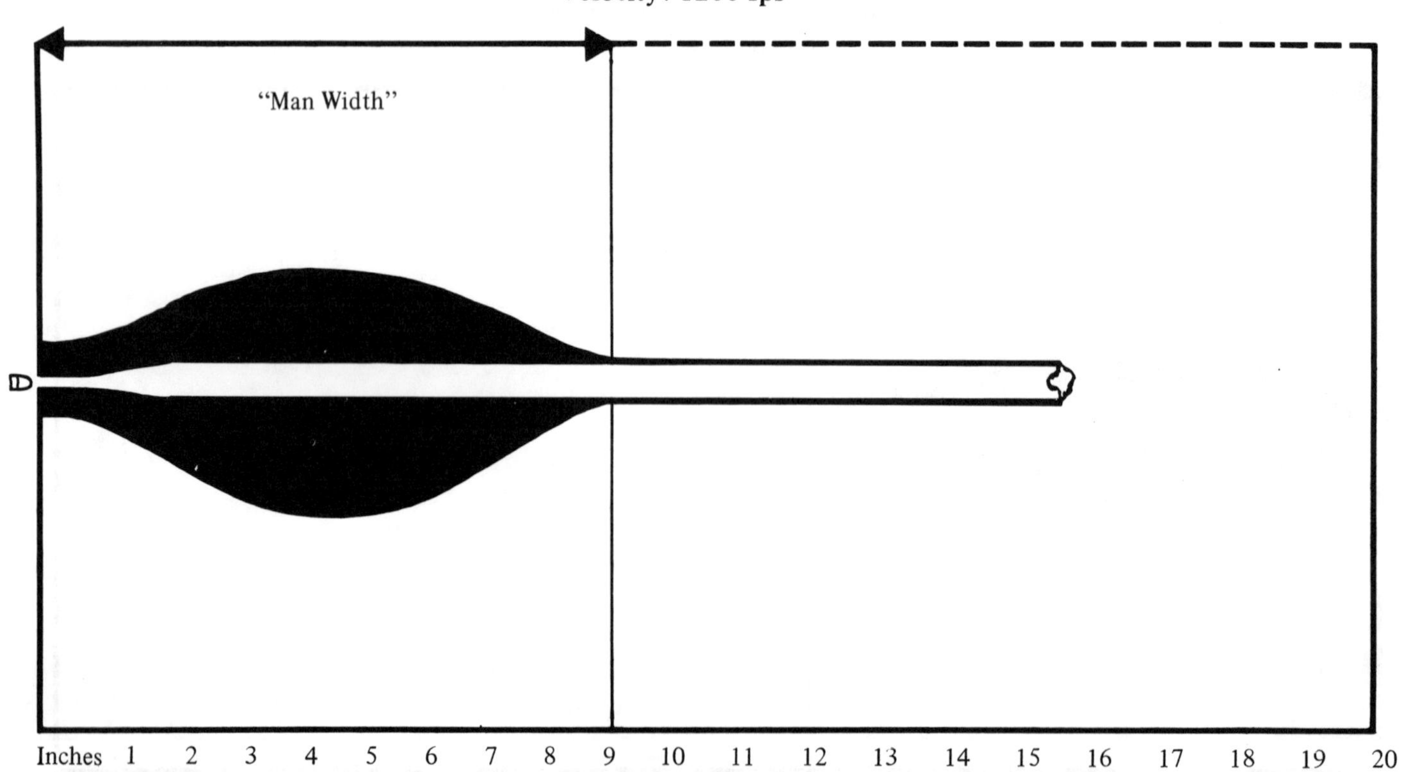

Permanent cavity moves down the center of the gelatin; the temporary cavity is shown in the shaded area. This bullet gives good penetration (though perhaps a bit too much) and the expansion needed in a combat round. Final diameter: 15mm (0.59-inch).

.357 MAGNUM PERFORMANCE IN BALLISTIC GELATIN

125-Grain, Jacketed Soft-Point Bullet

Velocity: 1400 fps

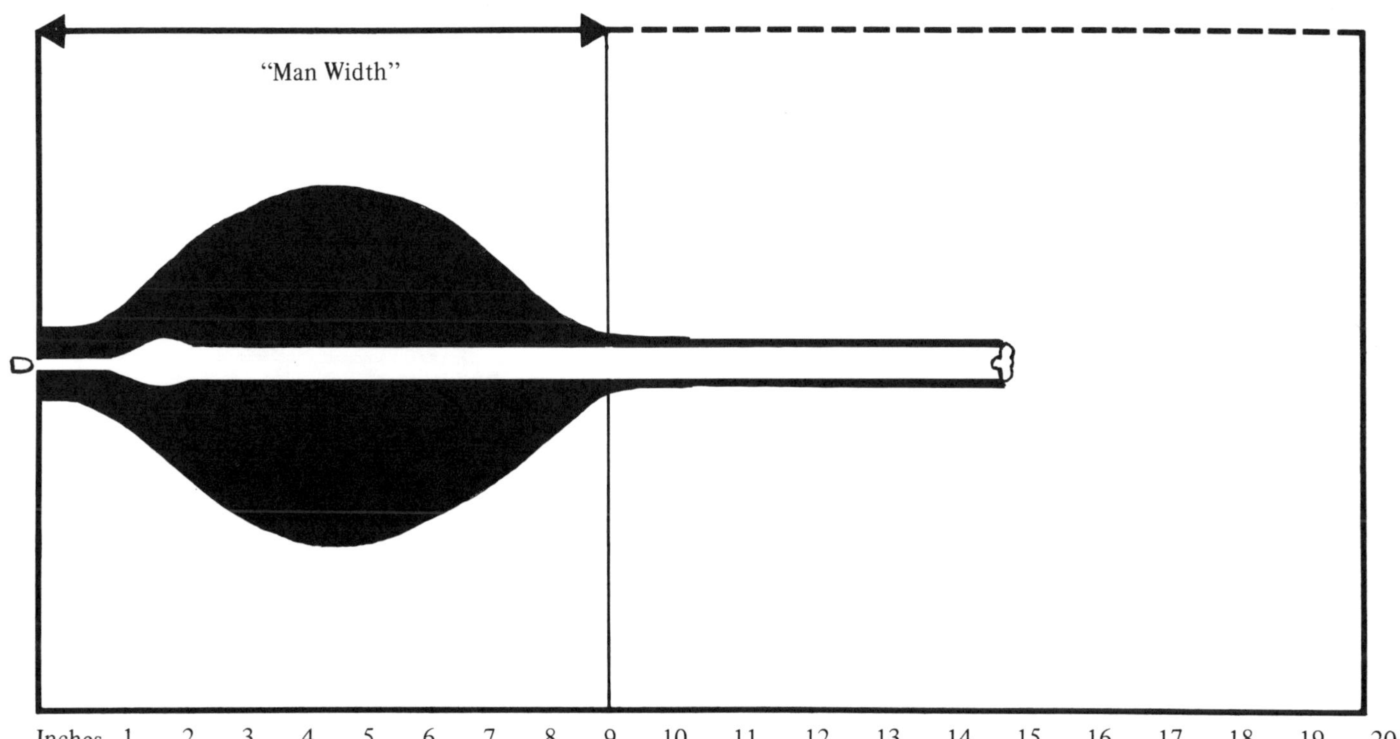

Permanent cavity moves down the center of the gelatin; the temporary cavity is shown in the shaded area. This bullet gives good penetration (though perhaps a bit too much) and the expansion needed in a combat round. Final diameter of bullet: 0.75-inch.